Pars Foundation
Lars Müller Publishers

FINDINGS ON ELASTICITY

Edited by
Hester Aardse
Astrid van Baalen

کا ارشاد ہے کہ جدید سب کچھ ہوتا ہے نصیب تمہاری ہے احمد ماما کی شادی

کی بات کا مطلب یہ ہے کہ تم اب بھی ترسے ہو کہ تم اپنا دل میرا نا چاہتے ہو

یہ کیسے ہو سکتا ہے ہماری اتنی پابندی میں اچھا خواب ہے تمہارا اور تم مجھ سے تو

اچھے خواب دیکھتے ہو میں تو وہ بھی نہیں دیکھ سکتی بس نصیب ہم دونوں

خود سوچتے ہیں کہ کیسے ملے عقیل کیسے اور بس بات کرنے کی ضرورت نہیں میں

بھیجنا تم بھی کچھ ضرورت دو تو میں بھی کچھ سوچتی ہر نصیب یہاں پر آئے تم اور

زیادہ یاد آتے ہو اپنا خیال رکھنا بہت زیادہ ہماری ہوتا

تاکہ ملنے کا ہر لمحہ مل جائے

اور کوشش کرنا کہ ۔۔۔ خدا حافظ

صرف آپ کی دیوانی

I love you s

اسلام علیکم

سلام کے بعد اپنے خدا سے آپ سب کی صحت چاہتی ہوں شکریہ اپکا دونوں بیمار بالکل ٹھیک
ہیں خدا کے فضل سے آپکو بہت فکر ان دونوں کو دیکھو پر اتنا افسوس ہوتا ہے لوگوں
بیماری کتنی عذاب میں ... دونوں جمعے ان کی بیماری سے اتنا ڈر لگتا ہے اچھا تم
کہاں ہے کہ میں مرد ہوں مجھے میں اتنی ہمت نہیں تو سمجھ میں تو بالکل نہیں آئی تم میں
اسکا عورت پھر تڑی تو بالکل کمزور ہوتی ہے تم کو کیوں ڈر لگتا ہے پتہ نہیں جمعے کہہ رہا
میں خود بھی اپنے بارے میں پرونت سوچتی ہوں کہ ایک دم جمعے کیا ہوگئے آخر کوئی
نہ ہوئی شعیب میرا دل بہت گھبراتا ہے ایسا لگتا ہے کہ ابھی میرا دم نکل
ہم اس بات سے مت کرنا ملنے کی خواہش وہ ہماری ساتھ بدنا ہوں
تم اُس کو خط بھی سے دینا کیونکہ سب شکل کرتے ہیں ہماری چیلی
پہلے دوراز سے ہیں جمعات خانے کی طرف اُس میں خط میں تم سے
تم اس کے بعد ... کے سیکلر ہی منٹ پر آپ پڑو میں اور پڑ سے پڑو
اور اُس بعد وہ اسے خط کودی شکل ہے کودی اور ہمارے
ہو جائے اس کے بعد میں ہم اس زیادہ نہیں ملوں گی اور تم بھی
مت جانا کرو خواہش انسان اپنی ساست خراب کرتا ہے شعیب
میں نہیں آنا کہ میں تم سے کیا بات کروں یہ نہیں شعیب بالکل بالکل
سمجھ میں نہیں آنا کہ آخر میں کرو کیا شعیب میں موت سے
مگر تم ہر وقت موت کی باتیں مت کیا کرو جمعے اچھا نہیں لگتا ہے
آنا بے غیرت ہو ونا ہے وہ زدگی موت کی بات کرتا ہے شعیب
کہ شادی کی بعد نگلو گیں گا اور تمہاری میرے ساتھ دوستی
لنے ہو تم دونوں شادی دی کہ بعد خوش رہو گے کیا یہ میں ہے محب
میں تمہاری ساتھ دوستی یہ پیدنے دور

نہیں دھوکہ گیا پہنچے ہے محب

CONTENTS

FOREWORD

For this second volume in the Atlas of Creative Thinking, Pars invited creative thinkers from around the world to share with us their findings on elasticity. These, ranging from quirky, humorous and beautiful, to mind-bogglingly complex and disturbing, are what make this book. Why elasticity? Well, we're surrounded by elasticity, whether it's in the shape of our skin, the price elasticity on the Stock Exchange that plays havoc with our spending power, or a piano string that stays in tune for as long as its tension remains constant. It also never ceases to amaze me how resilient people can be in the face of adversity; somehow we manage to pick ourselves up by the hem of our lives and bounce right back.

What actually is elasticity and how does it relate to our everyday lives? Take something as ostensibly simple as the elastic band. It has length, width and thickness. They make perfect slings and keep post packages together. I remember as a nine-year old being sprinted after by a surprisingly athletic pensioner, round the flowerbeds poshing up the car park in front of the service flat where my grandmother lived. I had aimed my sling at his window that stood invitingly open on the second floor, and with some success. He had happened to be sitting on the edge of his bed tying his shoelaces when the clump of earth came whizzing past, cut a path across his bedroom, flew down the hall and finally gunged the door of the fridge in the kitchen. When the old man, choking in equal measures on physical exhaustion and outright indignation, eventually related this rather remarkable trajectory to my grandmother, my initial fear quickly swelled into Amazonian pride. I can't remember the punishment I received or the gist of the lecture (there will almost certainly have been one), but I can clearly recall inwardly rejoicing at how my skills with the elastic band, which had been wrapped around a package delivered with the morning post, had been more successful than I could have ever wished for. I learned early on in life that elasticity can boost one's self-esteem enormously.

مورط نہ دینے ہو تو ہم دونوں

بہت قریب آ ون نزدیک بردو

ذخیرات کے بارے معلومات تھا ہو گئا

باری دوارے بارے کہ میں کہتا ہ قدیم اگر میں جلدی جلدی

آ سکتا ہوں ناراض مت ہوجاؤ تمہارے اگر جواب سب نے شاید

معلوم ہو چکا ہوگا۔ بلو تم بل فلرمت کرنا اتنا میں نہیں

اور ہور اتنا ہی میرا جحبت زیادہ ہوجایں۔ کیو یہ شادی

پہل بہت نزدیع احھا نہ دیکھاہ۔ قدریہ جھاتا ہو لدیچ صرف

بلو۔ قدریہ احھا نہ لکا یادہ نزدیع لا ایک سا نھ ہو

تمہار ہمراو کا بہت بندرہ نہ کہ ناپس میری بیجد یا کہ

را اردو لکھا اور یہ احھا ہو دائ ۔ میں صرف تمہارے خاطر اردو بولنا اور لکلنا

سا و نگاہ و دنہ مچ لسق ق نہ اور دیا ہو ودا

سراخط ہمارں جعا بیس نہ لکگہ مں نا یاد س لسا سوت

بے موخط لبلیا ہو پے یچ جس ایدنیاد کم صعرا اوا ٹم

ذاچہ ان صعد دارو محم خفا لیکیچ ہو قدریہ اسلام دریچ ہم میں اپنے آ بے

تو بہت خوبی قسمت سمچکا ہو کہ میں صرف اپنے پیا رواپنے مجت کو

خفا لیکنا ہو الاجب تحفہ تمہاری بیجد سا ہو تو تحفہ تو کتنا خو مرہا ہے

کہ یہ تحفہ صرف تمہارے آ ط میں آ تا ہے میں ہردن سوح سوح کر لمچ

خط لبلنا ہو ہ ذدریہ اور قدریہ تمہاری یادں محم بہت تنگ سدنا یا یہتے ہے اوس

وقت کم ناراض نہ محم حضرت نور کے رسانہ انک ملاح یا بس پیا راو دم

ہلاس نطف یا نہ ہم لیا ہو کیاہ میں ہدیوں ہردات لسے کی یاد ے دوب

کر رہا ہو ہر بل میرے سا نھ ہ ہر کام ہ میرے سا نھ ہ ایک نرر

یعنی بھلائی کاہ اور میں تمہارے لہ ضرورت میری جان الکو تمہارے کاوقت ہے طلار الکو کہے
مقصد نہ جانے اور الکہ عادت پید اہوا آیا تھا میری لیے مجھے دات بطلاح خواب آیا تھا اور
میں سوتا تھا اور جب رات کے ایک بجے یا دوبے ہو جاتا تھا تو میں بید ار ہو جاتا تھا اور دکھ
بید نہیں آیا تھا اور جب نیند آیا تھا تو میں خواب میں آیا تھا اور میں خواب میں تمہارا کام
بلند آواز لگاتا تھا نہ تھلایہ تکدریہ تکدریہ اور یقدر یہ حب میں الکہ وقت تکلو خواب میں
دیکھتا ہو تکلو تھا رے الکو کے سا تھ خواب میں دیکھتا ہو اور الکہ رات کو خواب دیکھا کہ
تو میرا بروی ہو تھا اد سرخوان میں صرف میرا الو اور اح ہے او صرف ہم الو دم ہے
الو دیم تھا نا میر اکہ داخلا رضیم تکدریہ میں بہت نہ ت اجھا خواب دیکھا ہو ہر دات
میری خواب میں ا سر وقت ترہو اور میں تمہا رے علاوہ کہ لو خواب میں نہت یا
دیکھتا ہو میں الیس تمہار ہا ہو اور الکہ رات خواب میں میں مجھے الکہ ڈسمال
اور کھلاتا نا میرا خواب نونت ہم اجھ خواب ہے خدا کرے اور دیم اجھا ہو جا یے
تمہارا دعا سے الو مقد دیم یعنی اپنے الکہ خواب ضرورت حط میں لینا نہ تیرا خواب
لیسا ہے لین خواب میں جھوٹا ہو لینا بہت تریدی ڈرا ہے میں الیس خواب
سیے لیکم تکدریہ دوقوف دار سو ا نی جو تو نہ دیں ہو کہے لو ان لو نو کا نا
یعنی بتایا جو میری باتی میں برا لگتا نا اور الو سی بات میں جو ہم ناراض نہ
نہ بتادیا نہیں لو تکدریہ کو میرا اعتبار نہ ح اور دوسرا خواب لوخراب ہو یے اس
کو ہمرے میں لکھنا ہو اسلام علیکم میری جان سلام کے بعد تکدریہ بہت یا دآیا
بہت زیارہ اور تمہاری عید بہت بہت مبارک ہو خدا کرے میں ہم دونوں
مل اجھا عید ہو دا اور تکدریہ میں اپنے خدا سے دعا کر نا ہو کرا کہ ہم ہو یے
کو تکدریہ کو یعنی موت دیدو اور تکدریہ ما ہو کہ طلی دیے ہم ہو نہ دیدو اور نہ

There was an elastic band wrapped around this book. Walk down any pavement with your head down and the chances are you will find one lying abandoned on the ground:

Oddly and mysteriously, a discarded elastic band will typically present itself in the shape of an infinity symbol (properly called a lemniscate). Perhaps that is because elasticity possesses that elusive quality of the infinite. Bound and unbound. Infinitely big or simply unimaginably big. It never ends. I am not sure how, but it can make the universe be infinite, forever. That takes the pressure off getting things right. Come to think of it, elasticity makes things possible. Elasticity has no inhibitions. Science has no inhibitions. It makes the infinite shamelessly possible and time irrelevant. Robert Hooke, the first to formulate the law of elasticity, describes this linear variation of tension in an anagram,

'ceiinosssttuv': 'Ut tensio, sic vis', meaning 'as the extension, so the force.'

Does this explain why I often feel I'm just going back and forth, and back and forth? Is this life's elastic force championing the outer limits of *my* extension? Millions of elastic bands scatter the streets but I can never find one when I need one.

As science continues to shamelessly stretch knowledge as far as it will go, unburdened by inhibitions, so art, in its limitless ways of expressing human experience, often confronts our inhibitions and suggests where we should put them. The back and forth bond between art and science, whether it be through technological advances, means of representation, shifting ethical outlooks, a shared curiosity or overlapping subject matter reminds me of an article in *The Independent* in which a robot at Reading University is reported to have replied, 'We live in eternity. So, yeah, no. We don't believe,' in answer to the question, 'If we shake hands, whose hand are you holding?'

In a world of ever-narrowing specialisation, and one wherein human experience sometimes feels as mixed up as mixed media, it is necessary to take time and stock and assemble the findings of these two generic groups—art and science. *Findings on Elasticity* offers the reader the opportunity to see, even to undergo how they inform each other, and equally important, how they invade and influence our everyday lives.

For the making of this book, guest advisors were asked to nominate artists and scientists whose work relates to elasticity and who shape the way we look at the world today. Rather than letting these creative thinkers return to their garret, laboratory or studio to write up their notes for specialist journals or show their work at temporary exhibitions, we asked them to share their most important, crazy, amusing, aggravating, confusing, contradictory musings or comments with us. Hester Aardse and I stipulated only two conditions: that their response pertain to elasticity and that it reflect the language of their profession. As a result *Findings on Elasticity* is a book about elasticity but equally about the ways in which artists and scientists describe the world.

A painting has length, breadth and context; an event consists of time, space and form. In both cases, the viewer, or as in the case of this book the reader, is its fourth dimension. It is in this fourth, elastic dimension where we come to grasp, (mis)construe and imagine. We have left much for the reader to discover; to make different analogies from ours and to be inspired.

Astrid van Baalen, March 2010

تمنئ کوئی کاراذازہ نئ دی مدتھاری ایک نوکو لیلو اور سا بھوتلوگ

تو سر سی لیلو یا وتگ یار دحوحبت ترجا جوحہ ...ما ... ملو

یبار دگا خوئی دگا ... ... تملو

I ... ... Jan

I love you very very very Much kabhin Jan

قدریہ قدریہ قدریہ قدریہ قدریہ قدریہ

قدریہ جان قدریہ زندگی قدریہ نگلمون کالود

قدریہ دل ... القدریہ جو قدریہ نی نظری

کم بعنہ موقلیہ دل ... دھری قدریہ مر

خوئش قدریہ شکیب ... ... قدریہ مراسب

... جود ... ... ای

جادولک ... نظر خوشبوکرالدل

نوح ... مرگ قدریہ جاک ... ن

قدریہ قدریہ قدریہ قدریہ قدریہ قدریہ قدریہ قدریہ

قدریہ قدریہ قدریہ قدریہ قدریہ قدریہ قدریہ قدریہ قدریہ

قدریہ قدریہ قدریہ قدریہ قدریہ قدریہ

قدریہ قدریہ قدریہ قدریہ قدریہ قدریہ

اگر تمھاری اجازت ہو تو بہو قدیرہ جان اتنا پیسہ جو بھائی صاحب مانگتا ہے
اوسکا اتنا ہوتا ہی کہ ہماری زمین سودا ہو جائی اور شادی تک پیسہ
برابر ہو جائی مگر ہمکو تو اتنا کام کرنا ہو تو ہماری خاطر کرنا ہی اور ہمیں بھی اپنی
اُنکو بھی ہمکو کہوہ دوبارہ اپنی زمین خرید لینا ہی اسلئے اجازت مانگتا
قدیرہ جان اگر تمھاری اجازت ہو پیسہ نہیں مانگتا تو اب ایک ہفتہ
ہماری منگنی بھی ہی نور گیا تھا جب تک خداجان مہربان ہی تم فکر
مت کرو خداجان ہماری ساتھ ہی اور خداجان ہماری مدد ضرور کریگا
کہ ہم ساری دنیا میں بھی کوئی کمی نہیں ہی۔ اور اگر قدیرہ جان اگر
لندن جانی میں کوئی مشکل پیدا ہو گیا یا خداجان کا نہیں ہوا تو ہم
بہت خوشی سی ہماری شادی کرینگی وہ کتنا دن ہوگا قدیرہ جان
کہ ہماری شادی کا دن ہو گیا اور شادی کی رات میرا ایک مشورہ
ہی کہ پہلی ہم دونوں وضو کرینگی اور دو رکعت نفل نماز پڑھینگی اور اللہ کرتی
ہیں کہ خداجان اپنی پہلی جو ہم سی گناہ ہوا ہو کہ خداجان معاف
کریں اور ہمارے ماں باپ کو بھی معاف کریں اور ہماری زندگی میں بھی
خدایا نیک ہم دونوں کیا اگر ہم بھی کہیں جہاں ہماری بھی
اور ہمکو کبھی اور کا محتاج نہ کریں اور ہمکو ہدایت کریں کہ ہم
ہمیشہ اپنی خداجان کا عبادت کریں اور خداجان ہمکو کی نیک
نیک کریں اور بہت دعا کریں کے بعد دعا کرتے ہی کہ خداجان ایک
صالح اولاد ہمکو نصیب کریں اور نماز اور دعا کے بعد ہم وضو کے
ساتھ نزدیک ہو جاتی کتنا مزہ ہو گیا قدیرہ جان خدا کریں
کہ میرا دل تمھارا پسند آئی ہو اور کیا لکھو تمھیں

SIX HUNDRED PAGES

'There is no room for love in Afghanistan,' said a young teenage girl to me one day as we sipped tea in the sitting room of her family's apartment in Kabul. She said it as if it were true and had been true for years, for as long as she could remember. And not in that moment, but in the twilight of that evening and for several years after, her remark caused me to reflect on the kind of space that love itself can consume. An endless space without dimension, like a sketch without charcoal or a raindrop without water—more space than even the glorious mountains of the Hindu Kush could ever take up. Yet in the tiny precipice of this Afghan girl's heart, where love and all of its beautiful unknowns should have blossomed, it didn't, it couldn't.

The love that I felt in Afghanistan was a luxury. It was a luxury because I was an outsider and could afford to let Afghanistan enter me in a way that allowed me to recognize the beauty in all of its harshness. And although this land and its strife often, almost everyday, brought me to feel defeat, loss and compassion, I always had a great fluffy cushion to land on—a cushion provided to me by the love of my family and the memories of a life monumentally different to what I was witness to there.

From the moment I landed in Kabul, that love could have gone in several directions. It could have rested on the landscape or the children or the poetry that existed in the tired sighs of the people. But I was unequivocally drawn—as one is to light in utter darkness—to Afghan women. For them I had passion and energy. For them my emotions had no boundaries. For them I gave in wholeheartedly in order to show them to others as I saw them for myself; the most intricately designed butterflies stripped of their wings.

And then one day, a surprise. A young man I had known brought to me a stack of letters. More than six hundred pages. It was a secret correspondence of love, one that allowed the imaginations of him and his love to wander, for it was only in those pages and in their dreams that they could walk together. To disclose their love would mean the end and perhaps worse. That day I realized that love existed in Afghanistan, not without risk but in a single glance, a certain tone, the shadow of a schoolyard.

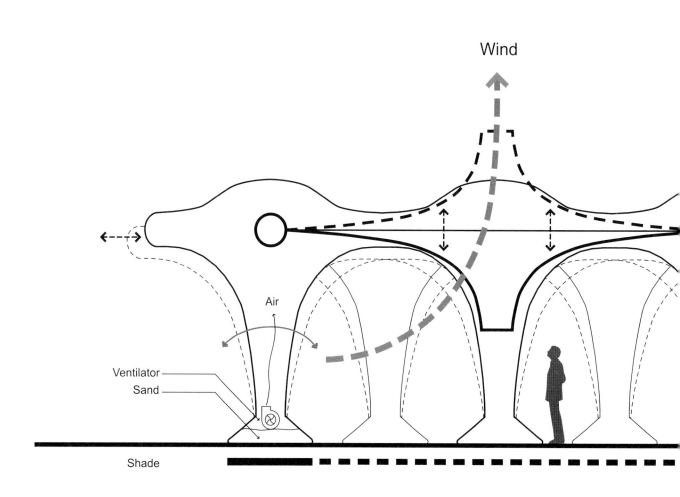

Wind

Air

Ventilator

Sand

Shade

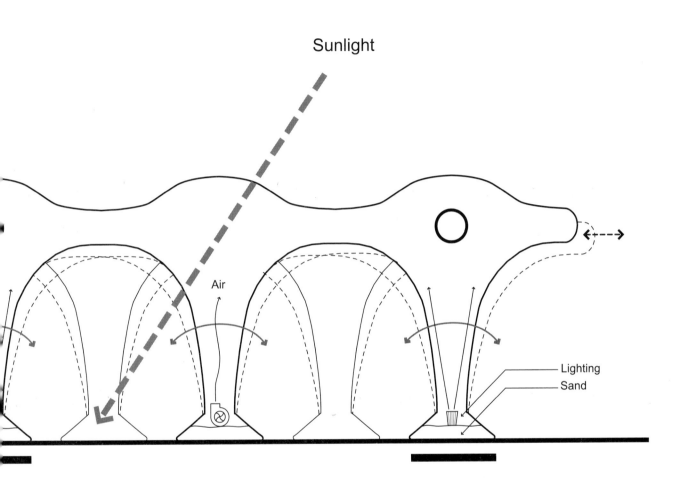

Sunlight

Air

Lighting

Sand

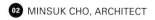

AIR FOREST

Air Forest is a 56.3m long, 25m wide pneumatic structure, composed of 9 hexagonal canopy units, at 4m height. These units are interconnected as one large piece of fabric, which are then inflated from the 14 blowers that are located at the base inside its 35 columns. These columns are 5m apart, and are weighed down by dirt and lighting elements that light up at night. The nylon cover is coated with a gradient of silver dots, whose reflective surface mimics the colours of its surrounding environment, as well as providing a playful dotted shadow on the people under the structure.

Air Forest is situated on the western edge of City Park in Denver. Being at one of the two disconnected open gaps of a ring of forests, this synthetic structure seems to be a continuation that bridges this gap.

The structure acts as a giant device to measure the site's conditions. Not only does it sway gently with the wind, it also acts as a barometer, since the installation becomes structurally weaker as the air pressure drops due to cooler weather, or even after sunset.

SHAPE YOUR BRAIN, SHAPE UP

Your brain is mutable and you can shape it yourself. This might come as a surprise, because many people believe that the shape and function of our brain is determined at birth, fixed throughout life and beyond our own control. However, this belief is incorrect. Our brain changes constantly. For example, the strands of a nerve cell that receive information from other cells (dendrites) branch and grow, as do the thick extensions of the cell that transport information from its cell to other cells (axons). New connections between cells are formed (synaptogenesis). The power of some of these connections is strengthened (Hebbian plasticity) while other connections are removed (pruning). Even new cells come into existence (neurogenesis). It is clear that our brain is not static, but dynamic. In brain terms, it is plastic.

This plasticity seems strongest when we are young but remains throughout our entire life. Plasticity can, within anatomical, physiological and genetic limits, be influenced in many different ways.

The knowledge that the brain is mutable can have far-reaching consequences for your life. It is ultimately to a large extent your brain that determines your emotions, abilities and behaviour. If the shape and function of your brain are not fixed, then your emotions, abilities and behaviour are not fixed either. If you can shape your brain, then you can shape yourself. But how do you shape your brain? The answer to this question is as astonishing as the question itself. You can shape your brain by your behaviour and by the environment to which you expose yourself.

What you do and what you experience moulds your brain. Painting, violence, hearing Chinese, writing, playing football or having siblings? Remem- bering, feeling cold or talking? Whatever you do or are exposed to, it shapes your brain because it influences your brain cells' connections and the communication between them. Through this shaping you develop your emotions, skills and behaviour. By deliberate shaping of your brain, you can develop these in the direction you desire, step by step in a continuous process of interaction between brain and behaviour. The development of your brain and behaviour is a continuous process of being and becoming.

You can make use of everything in this process: your motor skills and senses, but also your cognitive skills and emotions. They all exert an influence on the development of your brain and thus on the development of yourself.

There is currently a renewed flood of research that establishes how the specific use of your senses and motor skills leads to changes in your brain. It seems, for example, that in people whose arms have been amputated and who have gone on to specialise in painting with their feet, the motor areas in the brain that would normally be used for controlling the hands are now used to guide the feet. We also see this principle of transfer in patients with phantom limbs, in blind people who feel with their visual cortex, and in deaf people who see with their auditory cortex.

After training in a specific skill, the part of the cerebral cortex involved with that skill might increase. For example, the auditory cortex is 25% bigger in highly trained musicians than in people who have never touched an instrument. The size of the increase depends on the age at which the musician started training. It is therefore unlikely that these musicians already possessed a larger auditory cortex before they started their musical training. Instead, the training has so changed their brain that they commit more tissue to the skill. Therefore this skill can develop to a higher level.

In violinists we see also that the area in the somatosensory cortex related to the fingers playing the strings is larger than the area controlling the bow hand. The most commonly used fingers occupy the most space.

The finding that brain areas grow through training is interesting, but researchers conclude that the most important result of this has to be that the changes occur in all violinists, including those who started learning to play the violin at a later age. Even if you first start learning to play the violin at age 35 or 60, your brain will still undergo change and you will develop new skills.

This principle seems to apply to everyone and for all instruments that we learn to play or use.

Practice does make perfect and does so by changing the structure and function of your brain, step by step. This applies to all your skills, not just motor skills. Brain areas involved in the processing and storage of complex visual movements will grow if you learn to juggle, and learning to distinguish pitch will cause an increase in the activity in the auditory cortex.

Emotions also mould your brain. This has been most extensively explored for fear and depression but also for more positive emotions like happiness.

Cognitive skills, like concentration and memory, also reshape your brain. Eleanor Maguire and her colleagues showed for example that London taxi drivers have more grey matter in the back of the hippocampus (an area that is important for spatial memory) than those in the control group. It seems that the greater the driver's experience, the larger his/her hippocampus was. This suggests that the size of the hippocampus is affected by the use of your spatial memory. The increase in the back of the hippocampus was associated with a decrease in volume of the front part of the hippocampus.

Attention is crucial for our brain as well. Álvaro Pascual-Leone had a group of volunteers do five-finger piano exercises every day. Another group was required to merely imagine doing the same exercise. They played the exercise in their head, note for note, finger by finger. As expected, the real exercise induced changes in the volunteers' motor cortex, but what was truly striking was that *imagining doing the exercise (without actually moving a finger) induced the same effect, a change in the motor cortex.* The conclusion: the saying 'You become what you think' (so be aware of your thinking) must be taken literally.

Mental exercise induces changes in the motor areas in your brain. Consider for a moment the possibilities this presents. From behind your desk you can practise your backhand, or your penalty shot or Beethoven's Fifth Symphony, and actually get better at it.

DELIBERATE TRAINING

Physical exercise, using your senses, emotions and cognitive skills like thinking, remembering and concentrating have a direct effect on the function and form of your brain. This effect can be negative or positive. The more often you use something, or the greater the relevance of your experiences for your survival, the larger the impact on your brain.

Everything you have, everything you do, everything you think and everything you feel moulds your brain and thus your behaviour and your potential. From playing the violin to playing football, from feeling afraid to persevering, from distinguishing pitch to writer's block.

But we want to accomplish more than just shaping. We are interested in development. You want to mould your brain to develop your behaviour in the desired direction. But how do you do this?

There is no simple answer to this, but the expert theory of Professor Anders Ericsson, which promotes deliberate practice, can help us here. Deliberate practice consists of the right combination of sensory stimulation, physical activity, cognition and emotion. Composing training for a certain goal involves discovering which skills you have to master in all these areas in order to achieve that goal. In addition, you have to know which skills you currently possess and what limitations you have. (Naturally, there are anatomical, physiological and genetic limits that you have to consider.) Analysing these aspects is generally the job of a trainer or coach. If you know what you can do and what is needed, you know what you have to develop. If you want to learn to play the piano but you can't concentrate or have a limited range of feelings, then your training will have to concentrate on those areas, too, and not just on learning to read notes and doing finger exercises. If you learn to concentrate, you'll change your brain, and through these changes, other skills will become available like being able to practise one composition for an hour. The effectiveness of deliberate practice can be explained by the principles of brain plasticity. *While you practise your skills, you mould your brain by strengthening and pruning connections, by creating networks. By moulding your brain in this way, you develop new skills.* By developing your skills, you gain new possibilities for yet further development. Deliberate practice develops your skills and your brain, bit by bit, in a continuous process of interaction between brain and behaviour in a certain direction.

It goes without saying that if you set yourself an ambitious goal, mental or physical, you will have to work harder for it, talent or no talent. (What talent really means is that you already possess more of the skills needed to achieve the goal than most people. Your development will thus be faster.)

Optimising your skills is achieved by hours of concentrated practice and repetition which will fine tune your skills and lift them to a higher level.

In deliberate training the emphasis is not on routine or the enjoyment of the activity, but is fully directed towards change, thereby developing the cognitive, sensory, emotional and motor skills necessary to achieve the desired goal.

We have learned from Ericsson's research that there are significant similarities between the training of top performers in different areas. They practice for an hour with great concentration and then take a break. Napping is common. The sessions seem to be most effective in the early morning and a good concentration (and concentration training) is indispensable.

To reach a top level—in any area of expertise—you must train on average for ten years, seven days a week and four hours a day. Repetition, repetition, repetition... with attention for detail, which causes change. This is valid for even the most talented musicians, top athletes, artist and scientists. Research has shown that, even for those who are lucky enough to posses talent for something, those who have trained the most, perform the best.

If you are persistent and train in the correct manner it seems as if almost everyone, with a healthy set of brains, is capable of reaching a good level of functioning. Differently put: if you start playing the guitar (or whatever your dream may be) for four hours a day today, and you focus on those areas that you need to develop, in ten years you will be a meritable guitar player.

It is proven by children every day. Whether it involves eating with a spoon or putting on clothes, children repeat actions a thousand times to master these skills. Intensive practice still produces effects in adults because the brain remains plastic. Training and therapies prove it in abundance. The great thing is that as an independent adult, you are more capable of directing the effect of brain plasticity yourself. You can choose what you do and what you expose yourself to and therefore develop your brain in the desired direction. The effect of your choice is powerful.

Your brain is mutable. By shaping it you can develop towards who you want to be. However, we often don't give ourselves enough time, for whatever reason, to learn what we want to learn. We therefore become more of a user than an architect of our brain. We consume our life instead of producing it. That's fine, as long as you realise that you can also be an architect and a producer.

[excerpt and adapted from: *The Malleable Brain*]

References
Highlighted References, for full references see referencelist of *The Malleable Brain*:
Bosnyak, D. J., Eaton, R. A., & Roberts, L. E. (2004). Distributed auditory cortical representations are modified when non-musicians are trained at pitch discrimination with 40 Hz amplitude modulated tones. *Cereb.Cortex, 14,* 1088–1099.
Clifford, E. (2003). Neural Plasticity: Merzenich, Taub, and Greenough. HarvardBrain 16–20
Davidson, R. J., Putnam, K. M., & Larson, C. L. (2000). 'Dysfunction in the neural circuitry of emotion regulation—a possible prelude to violence.' *Science, 289,* 591–594.
Draganski, B., Gaser, C., Busch, V., Schuierer, G., Bogdahn, U., & May, A. (2004). 'Neuroplasticity: changes in grey matter induced by training.' *Nature, 427,* 311–312.
Elbert, T., Pantev, C., Wienbruch, C., Rockstroh, B., & Taub, E. (1995). Increased cortical representation of the fingers of the left hand in string players. *Science, 270,* 305–307.
Ericsson, K. A. (2006). The Influence of Experience and Deliberate Practice on the Development of Superior Expert Performance. In K.A. Ericsson, N. Charness, P. J. Feltovich, & R. R. Hoffman (Eds.), *The Cambridge Handbook of Expertise and Expert Performance* (pp. 683–704). New York: Cambridge University Press.
Eriksson, P.S. (2003). Neurogenesis and its implications for regeneration in the adult brain. *J.Rehabil.Med.,* 17–19.
Eriksson, P.S., Perfilieva, E., Bjork-Eriksson, T., Alborn, A.M., Nordborg, C., Peterson, D.A. et al. (1998). Neurogenesis in the adult human hippocampus. *Nat Med, 4,* 1313–1317.
Gage, F.H. (2003). 'Brain, Repair Yourself. Scientific American,3, 46–53.
Holloway, M. (2003). The mutable brain, *Scientific American, 3,* 78–85.
Gould, E. (2004). 'Stress, deprivation, and adult neurogenesis.' In M.S.Gazzaniga (Ed.), *The Cognitive Neurosciences III* (3 ed., pp. 139–148). Bradford Books.
Hariri, A. R., Mattay, V. S., Tessitore, A., Fera, F., & Weinberger, D. R. (2003). 'Neocortical modulation of the amygdala response to fearful stimuli.' *Biol.Psychiatry, 53,* 494–501.
Hill, N. M. & Schneider, W. (2006). 'Brain Changes in the Development of Expertise: Neuroanatomical and Neurophysiological Evidence about Skill-Based Adaptations.' In K.A.Ericsson, N. Charness, P. J. Feltovich, & R. R. Hoffman (Eds.), *The Cambridge Handbook of Expertise and Expert Performance* (pp. 653–682). New York: Cambridge University Press.
Karni, A., Meyer, G., Jezzard, P., Adams, M. M., Turner, R., & Ungerleider, L. G. (1995). 'Functional MRI evidence for adult motor cortex plasticity during motor skill learning.' *Nature, 377,* 155–158.
LeDoux, J. (2002). *Synaptic self: how our brains become who we are.* Penguin Books Inc.
Maguire, E. A., Frackowiak, R. S., & Frith, C. D. (1997). 'Recalling routes around London: activation of the right hippocampus in taxi drivers. *Journal of Neuroscience, 17,* 7103–7110.
Maguire, E. A., Gadian, D. G., Johnsrude, I. S., Good, C. D., Ashburner, J., Frackowiak, R. S. et al. (2000). Navigation-related structural change in the hippocampi of taxi drivers.' *Proc.Natl.Acad.Sci.U.S.A, 97,* 4398–4403.
Maguire, E. A., Spiers, H. J., Good, C. D., Hartley, T., Frackowiak, R. S., & Burgess, N. (2003). 'Navigation expertise and the human hippocampus: a structural brain imaging analysis.' *Hippocampus, 13,* 250–259.

so we have the
there is elasticity,

body, on the body

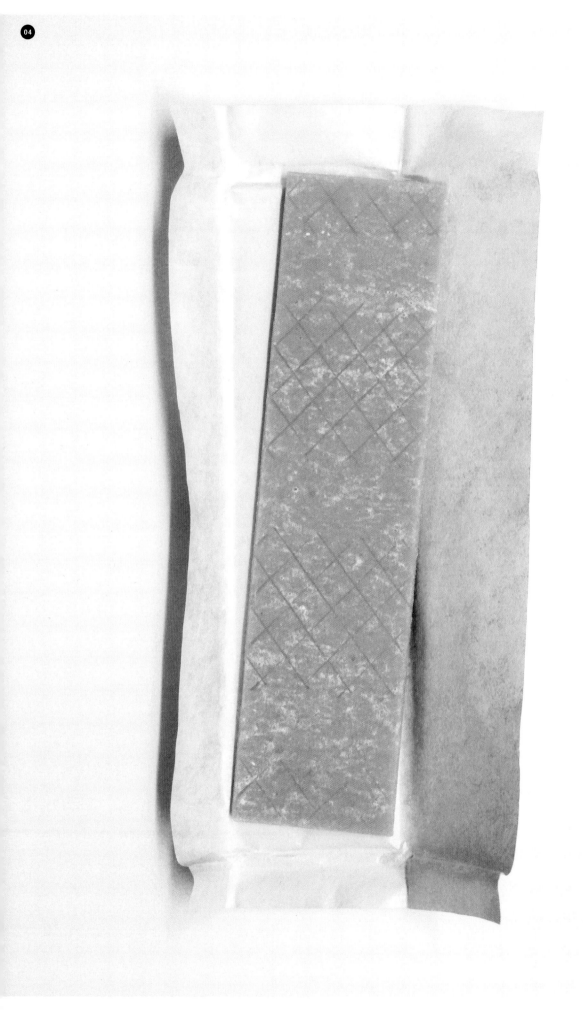

0'27"	0'53"	0'80"	1'06"	1'33"
1'59"	1'86"	2'12"	2'39"	2'65"
2'92"	3'18"	3'45"	3'71"	3'98"
4'24"	4'51"	4'77"	5'04"	5'30"
5'57"	5'83"	6'10"	6'36"	6'63"
6'89"	7'16"	7'42"	7'69"	7'95"
8'22"	8'48"	8'75"	9'01"	9'28"
9'54"	9'81"	10'07"	10'34"	10'60"

10'87"   11'13"   11'40"   11'66"   11'93"

12'19"   12'46"   12'72"   12'99"   13'25"

13'52"   13'78"   14'05"   14'31"   14'58"

14'84"   15'11"   15'37"   15'64"   15'90"

16'17"   16'43"   16'70"   16'96"   17'23"

17'49"   17'76"   18'02"   18'29"   18'55"

18'82"   19'08"   19'35"   19'61"   19'88″

20′14"   20'40"   20'76"   20'94"   21'26"

04

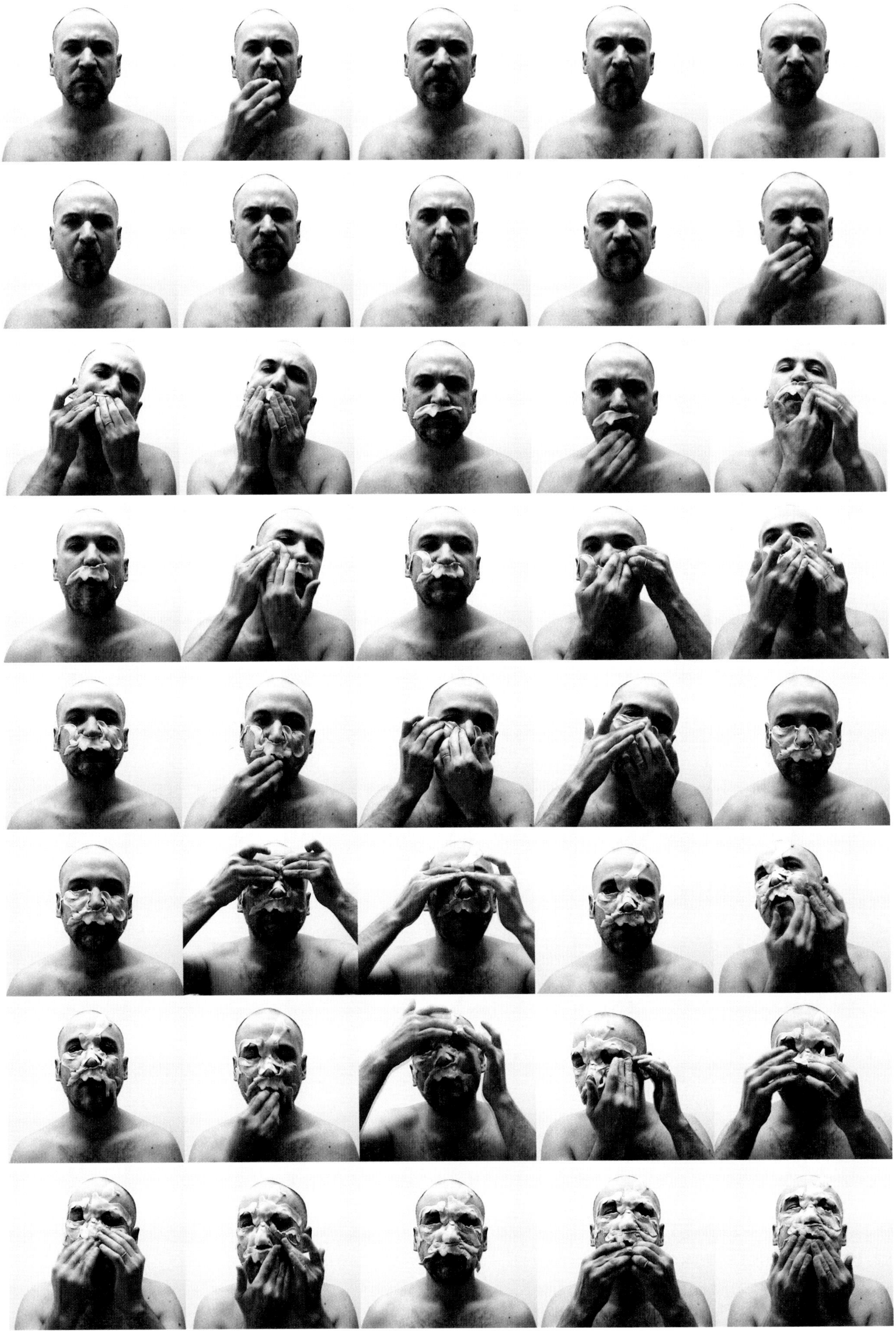

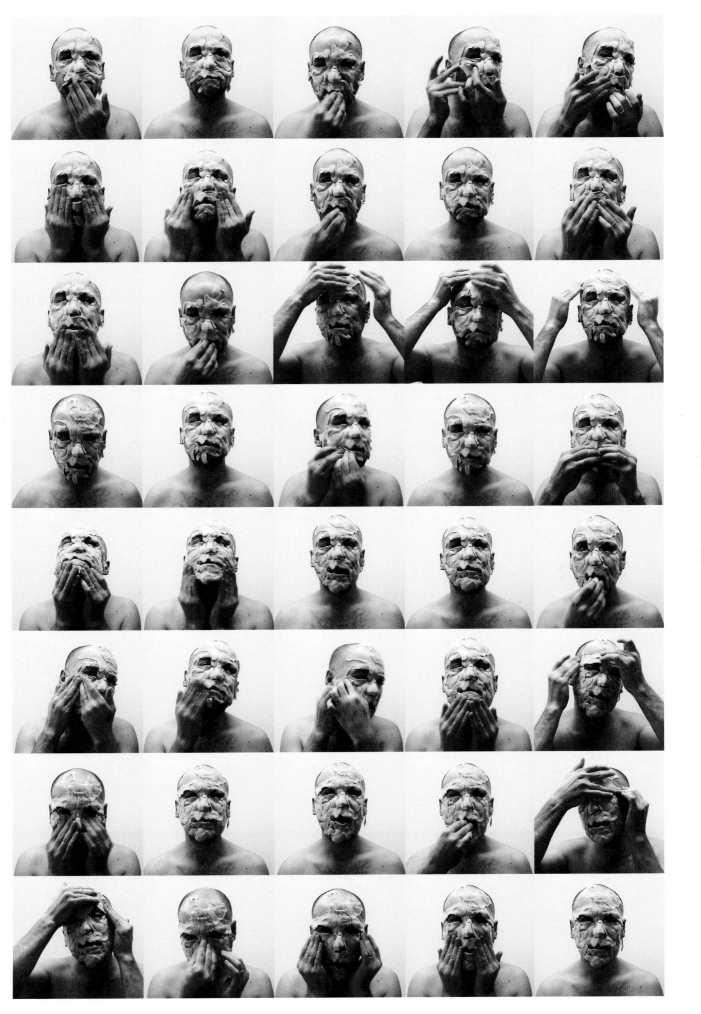

(G)HOST IN THE (S)HELL

Performance 21' 26"

STRETCHED AND FLEXED

This experimental sketch involves the manipulation of painted elastic and white fabric tubing. Because human hair has elasticity, Brooke, the model, pulled her hair with the elastic cords. While standing, she stretched the elastic and then released it. Brooke then pulled away from the tightly held elastic and experienced the uneasiness of the stretch.

Three-dimensional sketch, 2009
Nylon fabric tubing and painted elastic cord
Elastic cords: 149.9×14.9 cm
Nylon tubing: 14.9 cm wide
Model: Brooke Letourneau
Photo: Gary Pollmiller

THE ENIGMA OF THE LIZARD'S TAIL

My team and I have just had a business meeting about the characteristics of the textures that hydrocolloids bring to new dishes. These thickening agents are used to set, gel, and stabilise a variety of foods. Our discussion revolved around flexibility and the fact that it appears to appeal to the visual and tactile more than to our other senses of taste, scent and sound. Proteins and gums create gels and the resulting sensation, when the tension of these gels is both fragile and changeable, is one of pleasure.

Our meeting's over, and I set off, dropping a colleague at the station on the way. During the journey I cannot stop thinking about the tails of lizards and the mystery of why, if a tail is severed, a new one grows in its place. I wonder about the mechanism which makes this miracle possible and the devices which restrict this process to the tail and not, say, to the legs. Undoubtedly this question has puzzled thousands throughout time. I am sure that the scientific world has provided us with answers and analytical studies on the subject, but nonetheless I still need to think this through myself.

I mull it over and arrive at a conclusion I consider quite obvious: the tail is an important part of a lizard but is not a vital organ. There must be a time factor somewhere in this enigma. The energy required to reproduce an organ has somehow to be connected with a resources/capacity/time equation. This leads to the question of how many resources can be dedicated at a determined moment to an action and its successful outcome. Perhaps the reptile is not capable of reproducing an organ like the heart during that crucial phase which lasts from the moment the organ is damaged up to the moment of collapse when it stops functioning.

My thoughts then move on to the development and evolution of other organisms and I end up thinking about human beings. What, then, are those elements, and energy, which make it possible for a child to develop so rapidly and why does this cease at the adult stage? To allow an individual to deteriorate till death, after the considerable effort that has gone into learning all those instructions and the experiences and education which have been amassed throughout a lifetime, seems to be a waste. What could be nature's motive in not providing a repair system similar to that of a lizard, accumulating energy as powerful as that which occurs in the early period of human development?

A thought intuitively flashes through my mind: elasticity. Life's prime objective is perpetuation, and therefore more than reinforcing the continuity of what already exists, it endeavours, in a gesture of surprising intelligence, to adapt to any eventuality. We can achieve true adaptation simply by being in motion and by distorting ourselves in given external circumstances, by transforming ourselves, an attribute so elastic it allows us in each evolutionary link to acquire, by means of small and vital adjustments, a shape which is practically unique when compared to the preceding one. Life has therefore fortified those qualities that are elastic in preference to ones that are lasting, allowing us at each step to transform ourselves, to stretch ourselves to a point where we cease to be ourselves. This must be the answer!

I have arrived at *Mugaritz*. A restaurant is somewhat similar. Its environment and evolution are elastic so that they can keep tradiations alive, while also constructing the customs of the future. Without loss of identity.

DAVID 'NOODLES' AARONSON
Elx, Spain, March 2009

EIGHTIES SMURF PLASTIC LINES PUNK RAP FRESH / RADIO MAGNETIC
LATEX BUBBLICIOUS PINK JUMPSTAR PLANET RUBBER AIRMAX FLUO
ASTRO FLUID STRONG CITY VISION UNBREAKABLE LARGE BRIDGE
STRAW KUNGFU JELLYFISH FUNK FLAVOR LYNX EXTRA FOOD CURVES
TARZAN FLOW BIG FLEX TIRES NEW YORK CHEESECAKE SLIME GREEN
INFINITE MARVEL MOVEMENT STRETCH VELCRO JUNGLE FURIOUS
COSMOS Mr FANTASTIC TIME WHEELS BRAIN METRO SPACE ULTRA -

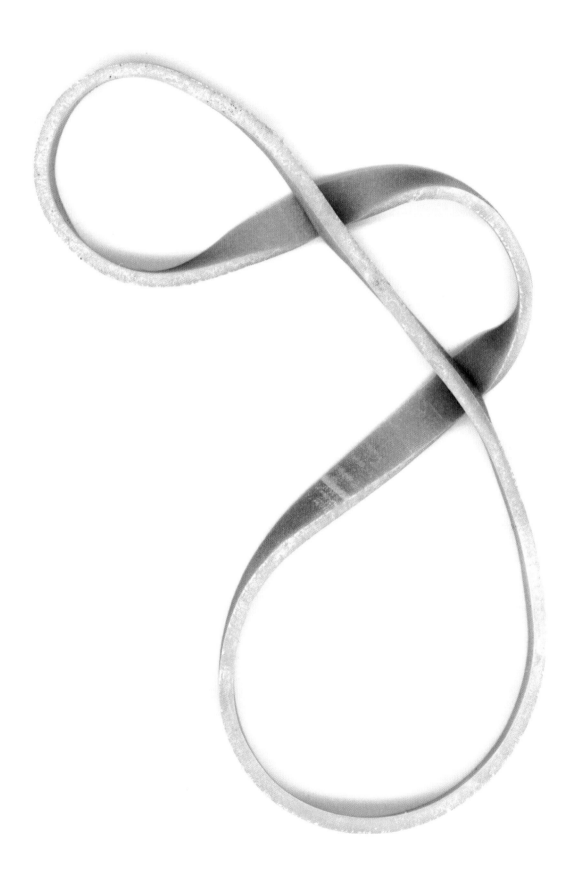

THE SEMI-LIVING

LIFE OF FRAGMENTS

How much can a body be fragmented before it is no longer a body? How much can life be stretched before it can no longer be called life? It is not death we are talking about here, but the semi-living. The artistic research project of the authors, The Tissue Culture and Art Project (TC&A), has been exploring and actualising these questions for more than a decade. We coined the term 'semi-living' to describe the type of in-between life forms that result from maintaining and growing parts of bodies in artificial conditions. By doing so, we try to engage with liminal life; with the notion that life can be stretched to places that greatly challenge our definitions and perceptions of what life is.[1] We explore ways in which the semi-living extend life to a point of no return. Although alive in some sense, and surely not dead yet, these entities are no longer part of the bodies and life that they once belonged to. Like slime mould, they might be able to reaggregate into a new kind of body or even migrate and rejoin some discrete bodies.

THE CELL AS THE BASIC LEVEL OF LIFE

Even though life seems to defy clear definition, an exercise in contemporary progressive reductionism will lead to the point at which death is no longer the outcome of expiry. This point, unsurprisingly, is most commonly referred to as the basic unit of life: the cell.

CELL THEORY

The idea of the cellular body dates back to Aristotle (340 BC) and Theophrastus (320 BC), who both described animals and plants as being made up of unified elements; blood and sap, flesh and fibre, nerves and veins, bone and wood. Later scientists such as Marcello Malpighi (1675) and Nehemiah Grew (1682) theorised that these elements are literally 'woven' into tissues of still finer elements.[2] In 1667, Robert Hooke, using one of the earliest microscopes, observed cell structures in a thin slice of cork. He coined the word 'cell' as the structure reminded him of a honeycomb. The second important development was the realisation that the cell, although part of a 'collective' body was in fact an autonomous agent, a 'little body' by itself. In making this claim, H G Wells and Julian Huxley argued that the term 'cell' was misleading, and they expressed their disapproval in their book, *The Science of Life: A Summary of Contemporary Knowledge about Life and its Possibilities* (1929):

Nothing could be farther from the reality. The proper word should be 'corpuscle' (little body) and not cell at all.[3]
We may compare the body to a community, and the cells to individuals of which this vast organized population is composed… It is very important to realize that this is not a merely allegorical comparison. It is a statement of proven fact, for—we resort here to the stress of italics—single cells can be isolated from the rest of the body, and kept alive.[4]

It was a botanist Matthias Schleiden (1838), and a zoologist, Theodor Schwann (1839), who were the first to formulate modern 'cell theory' as we now know it. Schwann wrote:

One can thus construct the following two hypotheses concerning the origin of organic phenomena such as growth: either this origin is a function of the organism as a whole—or growth does not take place by means of any force residing in the entire organism, but each elementary part possesses an individual force. We have seen that all organisms consist of essentially like parts, the cells; that these cells are formed and grow according to essentially the same laws; that these processes are thus everywhere the result of the same forces. If, therefore, we find that some of these elementary parts… are capable of being separated from the organism and of continuing to grow independently, we can conclude that each cell… would be capable of developing independently if only there be provided the external conditions under which it exists in the organism.[5]

ONE AS MANY, MANY AS ONE

While almost all multi-cellular organisms tend not to separate into discrete independent parts and rejoin again, and as a rule no complex organism can, there are a few exceptions. Among the most noted creatures that can easily shift from the unicellular to the multicellular is the slime mould. This odd life form is classified as a kind of fungus, which at different stages of its life cycle resembles unicellular animals such as protozoa, or a multi-cellular structure. The individual cells seem to flow into a higher form of organisation depending on external conditions. The idea that the same thing can be one and many is what makes these creatures so fascinating. But as soon as animals become more complex and their parts more specialised, this feat can no longer continue. With complex animals the idea of many discrete entities becoming as one was delegated to the social realm.

Due to modern day science and technological mediation, complex organisms can separate into many, and reintegrate into some sort of unified being as long as this process takes place outside the organism and in an artificial support mechanism.

1   A term used by Susan M. Squier in her book *Liminal Lives: Imagining the Human at the Frontiers of Biomedicine*, Durham NC: Duke University Press, 2004.
2   Ibid
3   Wells, Huxley & Wells, *The Science of Life*, p.26.
4   Ibid. p.27.
5   Theodore Schwann (1839) cited in White, *Cultivation of Animal and Plant Cells*, pp.188–190.

CELLULAR SENESCENCE

Most cells extracted from complex organisms usually have a finite life span, also called cellular senescence or the Hayflick limit.[6] Cells seem to go through degradation at the tip of their chromosomes, in an area called telomeres. After around 50 cell divisions these cells can no longer divide. Another issue with cell extraction is that when many cell types reach their ultimate function they cannot divide or reproduce any more—they go through terminal differentiation. It is believed that both cellular senescence and terminal differentiation are major contributing factors to the ageing of complex organisms. Research is currently under way to reverse the effect of ageing on a cellular level.

Some cells such as cancer cells, however, are considered to be immortal, in that they can theoretically keep on dividing forever. This can only happen once they are removed from the host body into artificial life support, as cancerous cells can kill the body that supports them. Telomerase, an enzyme that regenerates other telomeres at the tip of the chromosomes, has been used to try and make normal cells immortal.

Primary cells, the cells that are taken directly from an organism and have a limited lifespan, are increasingly seen as having the potential of reintroduction into bodies for the purpose of healing the organs or the body. This is the premise of regenerative medicine. Transdifferentiation is a process of reprogramming specialized cells and directing them to go down an alternative differentiation pathway, making new cells and organs from adult cells taken from the patient's own body. In a sense, the fluidity of life as manifested in the slime mould, in which cells can disintegrate and reintegrate, is being revisited through the more recent 'technologisation' of complex bodies and their parts.

THE SEMI-LIVING ARE NOT NEW

Parts of bodies have been sustained and grown, cultured, for more than a hundred years now. This is not 18th century Galvani-style reanimation, which relied on external, electrical charges, but the continuation of life processes and functions of parts that have been removed from bodies, be they organs, tissues or cells. Tentative attempts to keep body fragments alive were initially performed by Wilhelm Roux, who in 1885 kept embryonic chicken tissue alive for short periods of time. Ross G. Harrison grew a frog nerve cell outside of the body in 1907. In 1913 Harrison wrote:

… it seems rather surprising that recent work upon the survival of small pieces of tissue, and their growth and differentiation outside of the parent body, should have attracted so much attention, but we can account for it by the way the individuality of the organism as a whole overshadows in our minds the less obvious fact that each one of us may be resolved into myriads of cellular units with some definite structure and with autonomous powers.[7]

The first ongoing living existence of fragments—the semi-living—came about as a result of the more systematic and sometimes occultish practice of Alexis Carrel. Carrel cultured cells, tissues and later organs from 1913 to the 1940s. However it was not until 1948 that a continuous line of cells, originating from the one organism was established. The 'strain L mouse cell line' is still widely used in laboratories to this day.[8] Strain L was followed by the first continuous human cell line: the HeLa cells.

THE TECHNOSCIENTIFIC BODY

Maintaining and growing living fragments—the semi-living—involves the creation of a surrogate technological body (or epi-body) in which to grow the cells. This body —which we call technoscientific—provides the conditions that will allow the cells to grow and proliferate. In its most basic terms, this includes providing the right temperature, nutrients and other substances, and in some case substrates that promote cell growth. These fragments are unquestionably alive, satisfying at least the basic functions of a 'living' organism; they metabolise, grow and multiply. In the last couple of decades we have seen cases where the technoscientific body and the semi-living have formed a cyborgian entity in which function and feedback made it a responsive and effective unit.

Technoscientific bodies have been designed and built mainly for medical, pharmaceutical and military purposes. A small group has been building these technoscientific bodies for the purpose of philosophical and artistic reflection, observing and learning from the semi-living as evocative objects/subjects that are rapidly populating our human-made environment. Conversely, we are equally interested in registering the response of humans to the semi-living.

CELL FUSION

In some cases the cells of the semi-living fuse. Cell fusion is 'the nondestructive merging of the contents of two cells by artificial means, resulting in a heterokaryon that will reproduce genetically alike, multinucleated progeny for a few generations.'[9] When an undifferentiated stem cell fuses with a mature differentiated cell, the resultant cell retains the mature cell phenotype.[10]

Cell fusion among different species and different families along the evolutionary tree has been carried out successfully since the 1970s. Cultured Xenopus

---

6   In the early 1960's Leonard Hayflick and Paul Moorhead discovered that human cells derived from embryonic tissues can only divide a finite number of times in culture.

7   Ross Harrison, 'The life of tissues outside the organism from the embryological standpoint.' *Transactions of the Congress of American*

Physicians and Surgeons, 9 (1913) 63–75.

8   NCTC clone 929 (Connective tissue, mouse) Clone of strain L was derived in March, 1948. Strain L was one of the first cell strains to be established in continuous culture, and clone 929 was the first cloned strain developed. The parent L strain was derived from normal subcutaneous

areolar and adipose tissue of a 100-day-old male C3H/An mouse.

9   http://medical-dictionary.thefreedictionary.com/cell+fusion

10   http://www.medterms.com/script/main/art.asp?articlekey=32440.

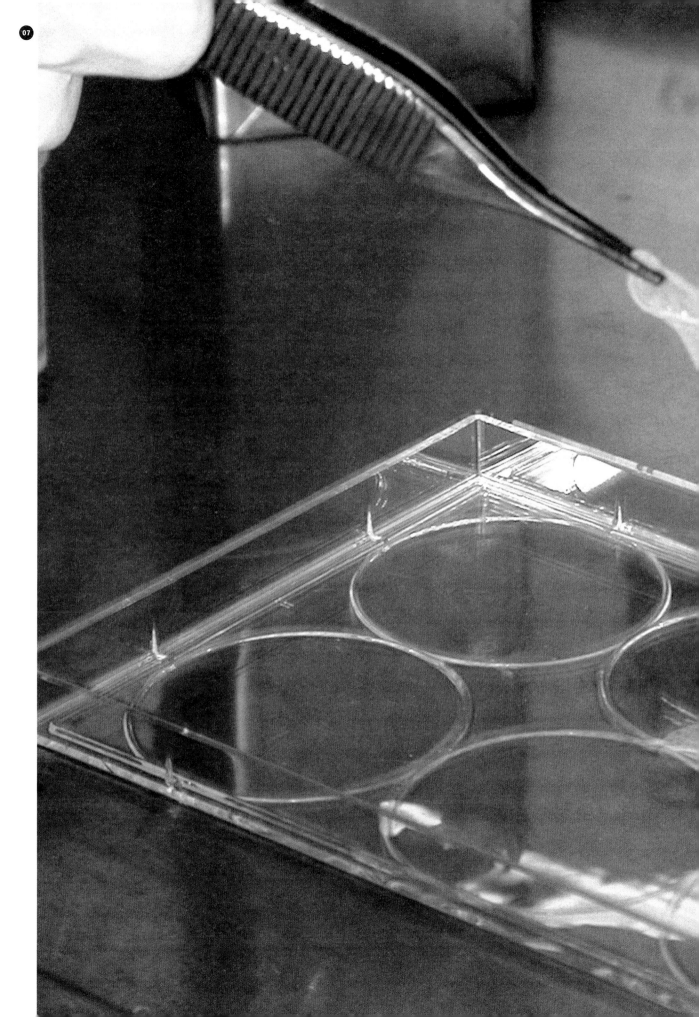

cells were successfully induced to fuse with carrot suspension cell protoplasts using PEG at high pH in the presence of high Ca2+. Ultrastructural observations confirm unambiguously that the fusion bodies seen by light microscopy are animal/plant cell heterokaryons.[11] The phenomenon of cell fusion, besides its practical applications such as a method for passing on specific genes to specific chromosomes, compelled Henry Harris to write about his experience as a pioneer in cell fusion techniques.[12] Harris' 2005 article opens with the following, somewhat romantic words:

There is a tendency for living things to join up, establish linkages, live inside each other, return to earlier arrangements, get along, whenever possible. This is the way of the world. The new phenomenon of cell fusion, a laboratory trick on which much of today's science of molecular genetics relies for its data, is the simplest and most spectacular symbol of the tendency. In a way, it is the most unbiologic of all phenomena, violating the most fundamental myths of the last century, for it denies the importance of specificity, integrity, and separateness in living things. Any cell—man, animal, fish, fowl, or insect—given the chance and under the right conditions, brought into contact with any other cell, however foreign, will fuse with it. Cytoplasm will flow easily from one to the other, the nuclei will combine, and it will become, for a time anyway, a single cell with two complete, alien genomes, ready to dance, ready to multiply. It is a Chimera, a Griffon, a Sphinx, a Ganesha, a Peruvian God, a Ch'i-lin, an omen of good fortune, a wish for the world.[13]

THE SEMI-LIVING IN ART

The semi-living, assembled from cells derived from different bodies, have the potential to become a 'new' type of body with 'new' functions. Our *Disembodied Cuisine Project*, for instance, explored the use of in vitro meat ('in vitro semi-living animal') for human consumption. Our piece *Pig Wings* was a symbolic comment on how functions change—these pig wings, of course, never flew.

 07 untitled

Description: A layer of pig mesenchymal cells (bone marrow stem cells) grown for 6 weeks. Grown as part of Oron Catts & Ionat Zurr residency as Research Fellows at the Tissue Engineering and Organ Fabrication Laboratory, MGH, Harvard Medical School 2000–2001.

The *Pig Wings Project* (2000) was a response to the hyperbolic rhetoric of biomedical research at the turn of the millennium; advances in bio-medical technologies such as tissue engineering, xenotransplantation, and genomics promised to render the living body a malleable mass. The rhetoric used by private and public developers as well as the media created public anticipation for less than realistic outcomes. The full effects of these powerful technologies on the body and society have only been superficially discussed. Winged bodies have been around in most cultures and throughout history. Usually, the winged creatures (chimeras) were perceived as either angelic (birdwing) or satanic (bat-wing). But there did exist a real vertebrate winged creature—far removed from any cultural values—the pterosaur. We have used tissue engineering and stem cell technologies to grow pig bone tissue in the shape of the three sets of wings the pterosaurs used to have. The Pig Wings installation presents the first ever wing-shaped objects grown using living pig tissue. This absurdist work presents some serious ethical questions regarding a near future in which animal organs will be transplanted into humans and semi-living objects will coexist with us.

In *Disembodied Cuisine Frog* skeletal muscle was grown over biopolymer for potential food consumption. A group of healthy frogs lived alongside as part of the installation. On the last day of the exhibition, the in vitro steak was cooked and eaten at a Nouvelle Cuisine style dinner, while the four 'real' frogs were released to a beautiful pond in the local botanical gardens. One complication that did arise from our victimless meat project was that it may have created the illusion of a victimless existence for the audience. In order to grow in vitro meat, a serum obtained from animals' blood plasma is needed with which to feed the tissue. Research to find an alternative is ongoing but with none in sight animals will keep on being sacrificed for the extraction of this serum.

In addition, growing in vitro meat is an extremely energy consuming and resource hungry process. But more importantly, this type of technological 'meat' production represents a shift from Tennyson's view of nature as "red in tooth and claw" to a more mediated nature.

The living are fragmented with parts taken away from the host body. These semi-living parts are introduced to a technological mediation that further 'abstracts' their inherent 'being aliveness'.

By creating a new class of semi-being, which is dependent on us for survival, we are also creating a new class for exploitation, as it further abstracts life and blurs the boundaries between the living and the non-living.

These new victims might be one further remove away from man and his conscience—they will still exist.

THE EXTENDED BODY

A rough estimate would put the biomass of living cells and tissues dissociated from the bodies that once hosted them in the thousands of tons. Then there are the tons of cells, tissues and organs currently kept in

11   M. R. Davey, R. H. Clothier, M. Balls and E. C. Cocking, An ultrastructural study of the fusion of cultured amphibian cells with higher plant protoplasts, *Protoplasma*, 96:1–2 (1978) 157–172.

12   Henry Harris, Roots: *Cell Fusion*, BioEssays, 2:4 (2005) 176–179.

13   Thomas Lewis, cited in Harris, *Roots: Cell fusion*.

suspended animation in cryogenic conditions, the maintenance of which requires intensive technological intervention to prevent transformation to a non-living state. This type of being (of semi-living) does not fall under current biological or even cultural classifications. The Tissue Culture and Art Project explores how the extended body may be classified and determined as a category of life in its own right. Much of this living biological matter can, in theory, be co-cultured, and fused (cell fusion), or share its sterile environment (to varying degrees of success). Age, gender, race, species and location do not play the same roles in the extended body as with other living bodies. This means that, in theory, every tissue in every living being has the potential to become part of a collection of living fragments brought together as an extended body.

However, whatever one decides the extended body actually is, and how independent it is as life, TC&A uses it symbolically as a unified body for disembodied living fragments, an ontological device that draws attention to the need for re-examining current taxonomies and hierarchical perceptions of life.

NOARK

NoArk, a recent artwork created by the authors, focuses on the taxonomical crisis this lineage presents.

Cell lines are catalogued and collected systematically in tissue banks, research institutes and patent offices. However, most of these systems have little connection to natural history taxonomy. This type of appearance of semi-life in the public arena is, say, more akin to the cabinets of curiosities then to the natural history museum. *NoArk* articulates this paradox and contrasts the natural history collection with the messiness of a small chimerical blob composed of modified living fragments extracted from a number of different organisms and made to exist in a techno-scientific body. In a sense, we are making a unified collection of unclassifiable sub-organisms.

By creating a device that allows the co-culturing and fusion of cells and tissues of different genotypes and phenotypes, ie from different organisms and different tissue types, *NoArk* presents the breakdown of both Linnaean taxonomy and molecular systematics (chemotaxonomy) and raises the question of semi-artificial life. A new technologically mediated ecology of semi-living fragments questions deep-rooted perceptions of life and highlights the need for the re-evaluation of our relationships with the living world around us.

The piece includes a fully functioning technoscientific body—a bioreactor and collection of (semi)living fragments of different bodies along with a collection of dead and preserved animals.

SEMI-LIVING WORRY DOLLS

In *The Worry Dolls* installation at the Ars Electronica Festival (Linz, Austria) in 2000, TC&A exhibited semi-living art for the first time. We grew seven semi-living worry dolls drawing on Guatemalan worry dolls. These tiny hand made fabric dolls are believed to take worries away. They are commonly crafted by small (children's) hands in the developing world and widely distributed to the more affluent countries. TC&A created a semi-living version made out of biopolymers and cells.

08 **Semi-Living Dolls Display**
Medium: McCoy Cell line, Biodegradable/bioabsorbable Polymers and Surgical Sutures.
Dimension of original: mix
Date: 2000

09 **A Semi-Living Worry Doll H**
Medium: McCoy Cell line, Biodegradable/ bioabsorbable Polymers and Surgical Sutures.
Dimension of original: 2cm × 1.5cm × 1cm

Our semi-living worry dolls might become our 'natural-ish' companions, invading and replacing our constructed and manufactured environments with growing, moving, soft, moist, and care-needing things. For us, the semi-living worry dolls can be seen as entities representing the associated concerns or 'worries' about our ever-increasing ability to further instrumentalise life.

THE BODY ELASTIC

The semi-living and the extended body could signal the erosion of life as it is broken down further and further to its elementary parts, until finally reducing it to its raw material. At the same time it can be perceived as a unification of life at the cellular level, a level where perceived borders have hitherto been considered near irrelevant. Through the dichotomy of our perception of the body as a whole and the body as discrete entities we wish to establish how elastic life is when viewed from the cellular and tissue perspective. Even when the whole body breaks into pieces, life rebounds and exists in some surprising configurations. Lab-grown life is more than just something that should interest scientists—it is raising some fundamental questions regarding how we see and how we treat living bodies.

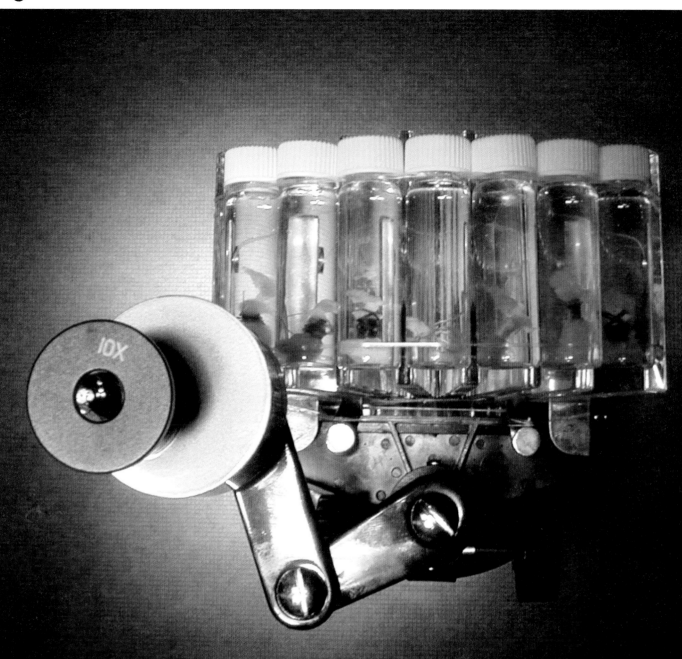

08

Our Product Range

We manufacture and supply all kinds of elastic, both standard and specialised:

- ▶ Woven elastic (including belt and brace qualities)
- ▶ Knitted elastic
- ▶ Braided elastic
- ▶ Round elastic (shock cord)
- ▶ Polypropylene elastic (skeleton)
- ▶ Corded elastic
- ▶ Finished/starched elastic
- ▶ Coloured and printed elastic
- ▶ Plush back elastic
- ▶ Gripper elastic

Other kinds of narrow fabrics, such as:

- ▶ Rubber tape
- ▶ Hanging/stay tape
- ▶ Cotton tape
- ▶ Herringbone tape
- ▶ Polypropylene webbing
- ▶ Shoelace tape

Our elastics are extensively used for:

- ▶ Pram covers
- ▶ Novelty hats and masks
- ▶ Shoes and ballet slippers
- ▶ Braces
- ▶ Belts (eg. nurses belts)
- ▶ Jackets (eg. cuffs, waistbands, etc.)
- ▶ Sportswear (eg. waistbands for jogging pants, gripper for cycle shorts, swimwear, sports fashion wear, and many other kinds)
- ▶ Ladies' casual wear (eg. waistbands for skirts and dresses)
- ▶ Men's casual wear
- ▶ Dance wear
- ▶ Protective clothing (eg. waistbands and cuffs for boiler suits…)
- ▶ Children's clothing
- ▶ Lingerie (eg. bra straps, knicker elastics, etc.) and more general underwear
- ▶ Men's underwear (eg. boxer shorts, etc.)
- ▶ Suitcase straps
- ▶ Ironing board covers
- ▶ Fitted bedsheets
- ▶ Specialised items, such as sponge bags and even for keeping open repair hatches on aeroplanes!

THE CONCEPT OF ELASTICITY IN ECONOMICS

When an economist wants to express how people react to a change in price, or a change in taxes, or a change in some other economic variable, he or she will inevitably turn to the notion of *elasticity*. Suppose, for example, that when the price of milk goes up by 10%, people respond by buying 5% less milk. Then the economist says that the price elasticity of demand for milk is 5%/10% = .5. The extremely useful property about this way of describing reactions is that no matter how we measure price (in dollars, euros, yen, or whatever) and no matter how we measure milk (in litres, gallons, pints, etc.), the elasticity will always be the same. That is, it is a *pure number*, without dimensions or units. This enables economists to compare how people react to changes in milk prices across countries and across time—making the concept of elasticity very powerful indeed.

MORPHOGENESIS

Generally, we are taught there are two predominant views that rule architectural form: one takes the idealist stance, where architecture is produced by a mind or an archetype that imposes ideas on inert matter; the other, materialist stance, states that architecture is nothing but the adding of bricks or steel beams to one another and should be equated with building. Both may be dismissed forthwith.

I'd like to leave the idealist stance to one side without any further consideration. The materialist stance is 'closer to the bone', but it is ultimately a mistake to think the matter of the building equates its architecture.

Building is about structure; architecture is about organization. Gottfried Semper, for instance, the 19th-century architectural theoretician, related the techniques of interconnecting materials by crafts then extinct in architecture to the driving organizational principles of design. His views on how textile techniques of weaving and plaiting influenced architectural ornament and the concept of the wall have been extremely influential. Then in the middle of the last century, Frei Otto used flexible materials such as sand, paper and even soap to calculate forms in a technique that can only be described as analog computing.

Combined or brought together under special conditions, materials interact and their agency results in forms that can subsequently be scaled up and built into another material. A pyramid is not very different from a pile of sand. Otto recalculated the famous shapes of the Munich Olympia stadium on a much smaller scale with soap film, resulting—what he called 'form finding'—in very precise patterns of minimal surfaces which could subsequently be built in steel.

The images shown here also relate two materials to each other, one informing the other. Materials like bone, or better, cancellous (spongy) bone as found in mammal jaws and leg bones, have very 'Ottoesque' properties in the sense that their structure is not genetically produced by the switching of chemicals on or off, but by pure, physical, morphogenesis. The complex foam-like structure emerges in a material process that constantly empties and lightens itself by formation of four-legged nodes that are much more stable than typical architectural nodes of posts and beams. A material that is wet and flexible constructs a system that is extremely rigid and structurally far more stable than most building structures. So, flexible, elastic organization becomes rigid structure. We are not aiming for an architecture that mimics organic, natural or flexible patterns—no, we simply use the same generative properties to transfer information of mud and foam to a world of glass, steel and concrete.

10

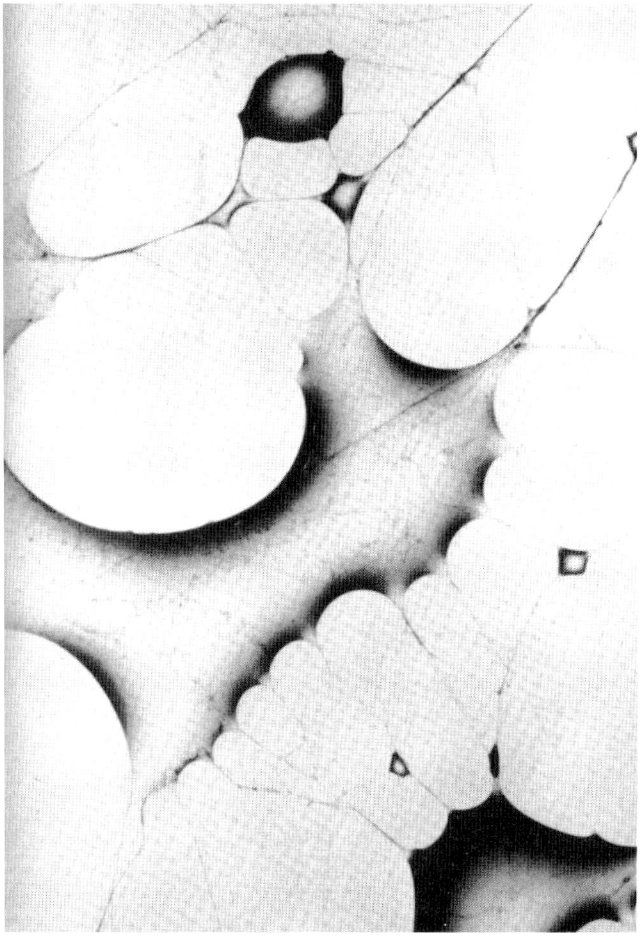

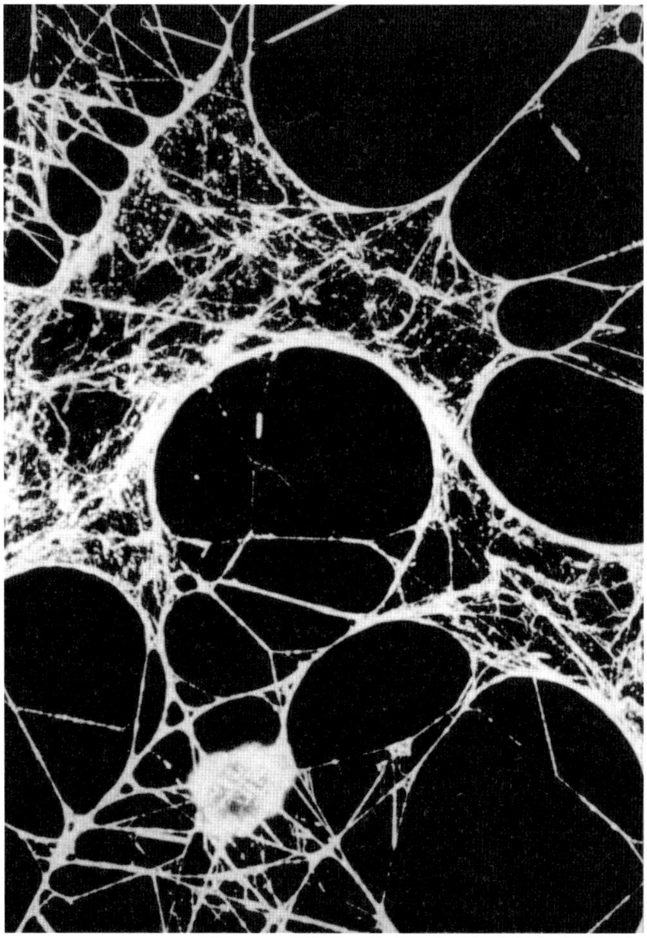

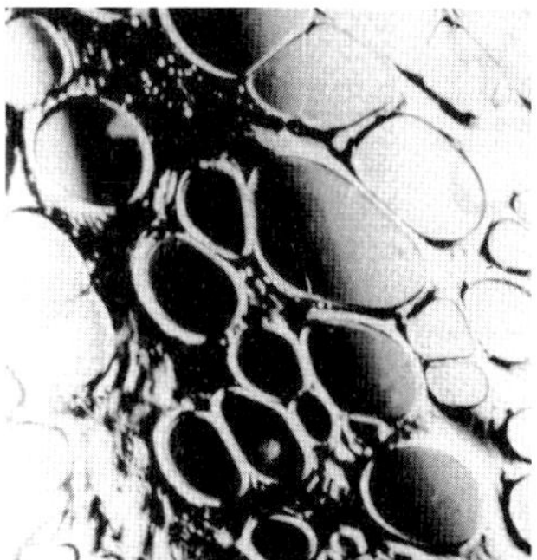

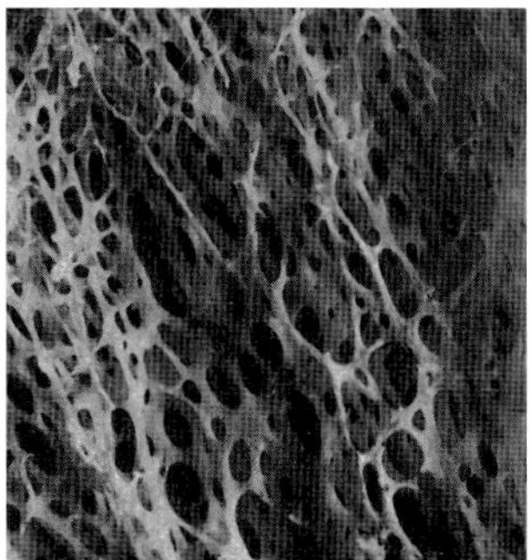

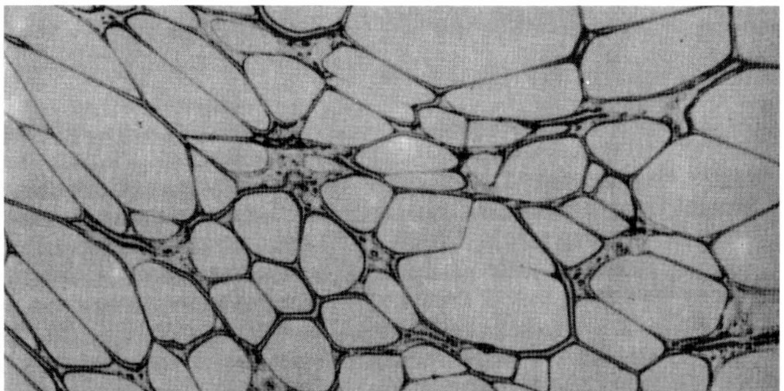

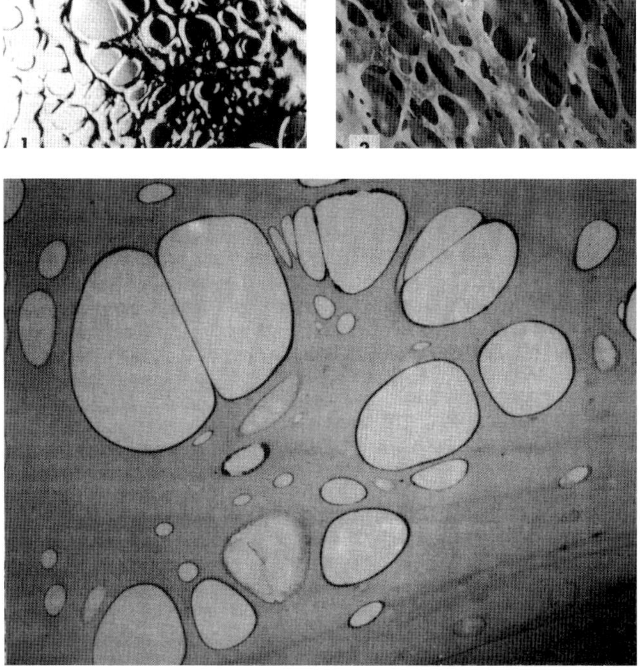

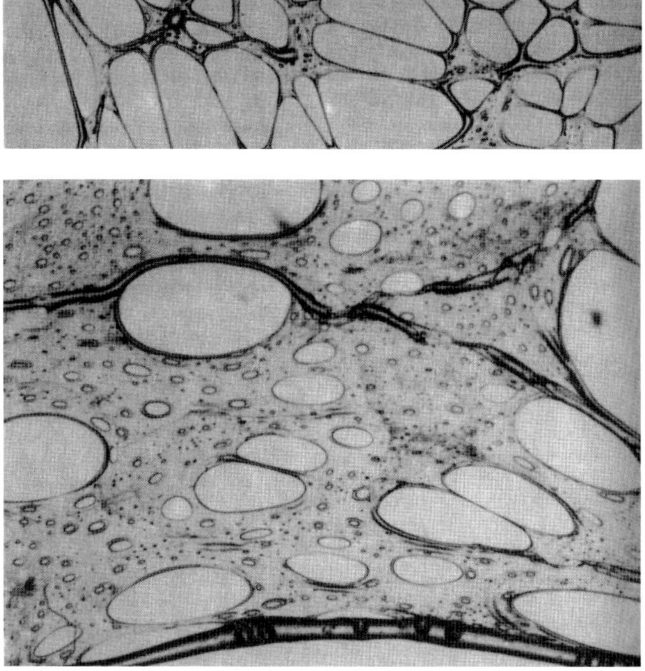

Lemma   Let $\theta(s)$ be a solution of the EL-eqn's (with or without contact) with $\theta(s) \to 0$ as $s \to \infty$. Then $p \leq 0$.

Proof   EL eq can be written as

$$-\frac{1}{2}\theta'^2 + \frac{1}{2r^2}\sin^4\theta - \frac{M}{r}\sin\theta(1-\cos\theta) - p(1-\cos\theta) = 0.$$

Close to $\theta = 0$ this gives

$$-\frac{1}{2}\theta'^2 + \frac{1}{2r^2}\theta^4 - \frac{M}{2r}\theta^3 - \frac{p}{2}\theta^2 \cong 0$$

If $p > 0$, then to highest order $-\frac{1}{2}\theta'^2 - \frac{p}{2}\theta^2 \cong 0$ $\lightning$.

$\square$

$\langle GRIN \rangle$

A PAGE FROM MY ARCHIVE, CA. 2002

It was the word <GRIN> at the bottom of the page that caught my attention. Through seven years of other work and unrelated thoughts this exclamation conveyed to me pleasure, pride and hope. Because for a mathematician the proving of a theorem—called a lemma in this case—is many things at once: a tool of the trade, a mark of craftsmanship, an intellectual achievement, a step forward, the answering of a question and, most of all, the opening of doors that lead further along the path of science. The topic is the theory of elastic rods, in this case rods that wind around a cylinder.

[image elastic rod around cylinder]

You might ask why we study such a strange setup. The long answer is too long; the short answer is 'start with simple problems, then move on to the more complicated ones'. In mathematical terms, a rod around a cylinder is a simpler problem than a 'free rod', as these examples suggest:

[6 images of free rods]

The lemma states a property about elastic rods, or to be precise, about models of elastic rods. The Euler-Langrange-equation

$$-\frac{1}{2}\theta'^2 + \frac{1}{2r^2}\sin^4\theta - \frac{M}{r}\sin\theta\,(1-\cos\theta) - p\,(1-\cos\theta) = 0$$

is a differential equation that any rod at rest should satisfy. The lemma states that p, the compressive force applied to the rod ends, can only be negative or zero. In other words; if one pushes the ends of a rod-wrapped-around-a-cylinder towards each other, the rod will collapse onto itself. (Note that one should keep the rod in contact with the surface of the cylinder all the time during such an experiment. Some things are easier in mathematics than in real life).
Clearly I was happy with this little result. Going back through the files of that time I can no longer reconstruct why. Over the years the property proved in this lemma has worked itself into my intuition, so much so that now I can hardly imagine ever doubting it. I should add a property to the list above: the proving of a theorem also modifies the intuition of the mathematician, and in that way shapes the development of the scientist as a human being.

11

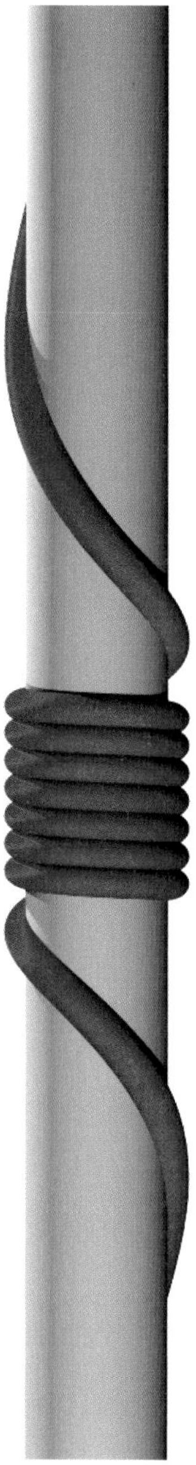

12

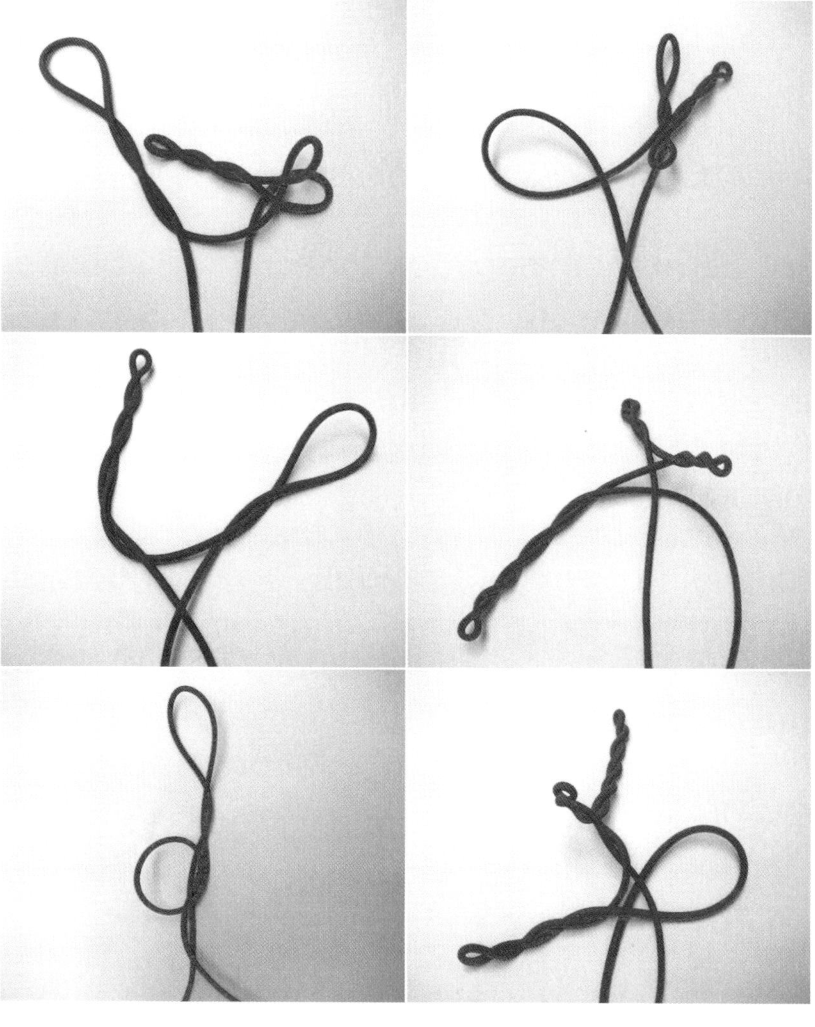

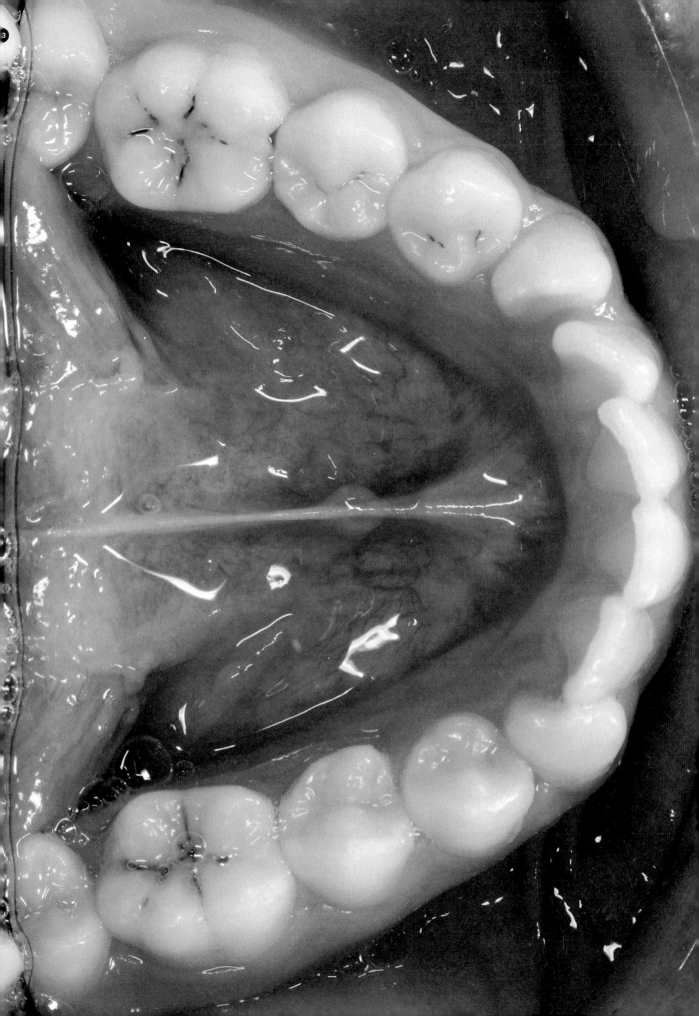

STRESS + STRAIN

Elasticity in orthodontics relates to the materials used for the exertion of force on the tooth, and the result of this force on the periodontal ligament and bone. This entails a wide spectrum of polymers and alloys used in orthodontic mechanotherapy, ie, the configuration of the applied force to provide the desired spatial placement of tooth, crown and root. Elastomers are utilized to retract teeth, and wires are inserted into brackets to place the tooth in a specific position.

ELASTOMERS; RETRACTING TEETH

The elastic behaviour of elastomers is the result of the reduced entropy normally associated with the distortion of a macro-molecular chain from its most likely spatial arrangement. The stress-strain diagram for this type of material shows the initial region of elastic deformation with a curvature, rather then being a straight line as for metals and ceramics. The elastic limit occurs at the point of inflection where the curvature of the plot changes from concave-down to concave-up. The area of elastic deformation corresponds to reversible changes in the separation between molecules in the polymeric structure. The absence of a linear region for Hookean elasticity in the stress-strain curve for an elastomer occurs because of the varying chain lengths and structures, whereas permanent deformation occurs from irreversible displacements between the polymeric chains.

Elastomers in extension are assumed to work as a network of flexible, irregular, yet homogenous and isotropic chains representing the mass of polymer chains. At equilibrium, when inward and outward forces are equal, it can be shown that stress is related to strain. When small strains are induced with a λ ratio (L/Lo) of less than 1.1, the stress at constant strain decreases with increasing temperature, while at larger ratios the stress increases with increasing temperature. This process is termed thermo-elastic inversion.

In clinical practice, the elastomers are used to ligate the wire into the bracket (braces). Therefore the feeling of numbness or mild pain, after each orthodontic appointment, derives from the replacement of the relaxed elastics with new ones, which show higher initial stiffness. This initial force will lessen over time.

SUPERELASTIC WIRES; PLACING THE TOOTH IN POSITION

The pioneer for the development of the super elastic nickel-titanium (NiTi) wires for orthodontics was G. Andreasen in 1971. The NiTi orthodontic wire offers a modulus of elasticity approximately one-fourth to one-fifth to that of stainless steel wires, along with a very wide elastic working range.

There are two major NiTi phases in the nickel-titanium wires. Austenitic NiTi has an ordered body-centred cubic structure, which forms at high temperatures and low stresses. Martensitic NiTi has a distorted monoclinic, triclinic or hexagonal structure, which forms at low temperatures and high stresses. The shape-memory effect (ie. its *elasticity*) creates a rapid martensite to austenite transformation through crystallographic twinning at the atomic level.

By inserting the NiTi in the low modulus form, the decreased stiffness of the wire allows for the engagement of the wire into the braces of crowded, skewed teeth. Upon exposure to the increased oral temperature, the wire experiences a phase transformation and is turned to the high modulus phase. As a result, forces start to bring the teeth into the desired straight arch form.

TISSUES; THE ELASTICITY OF THE PERIODONTAL LIGAMENT AND THE BONE

The application of forces on teeth results in stresses being distributed to the root and the adjacent tissues; these involve the periodontal ligament and the bone. Extensive research on soft-tissue elasticity has determined that the response of the ligament is composition-driven and varies with the mode of force. Application of tension results in compression on the upper portion of the root and tension in the opposite lower portion of the root. Compression of the ligament results in the distribution of force in the tissue and absorption of the applied load. This changes the shape of the tissue. Deformation of the membrane and nucleus of cells give way to bone reabsorption. Bone that is reabsorbed allows tooth movement and new bone grows after the stimulus ceases.

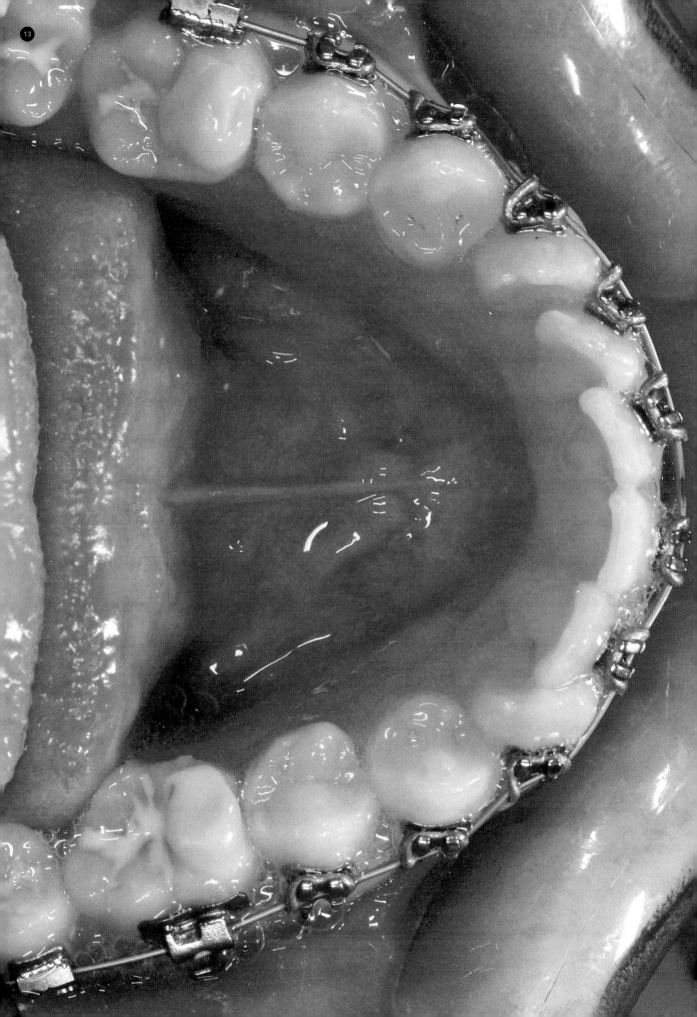

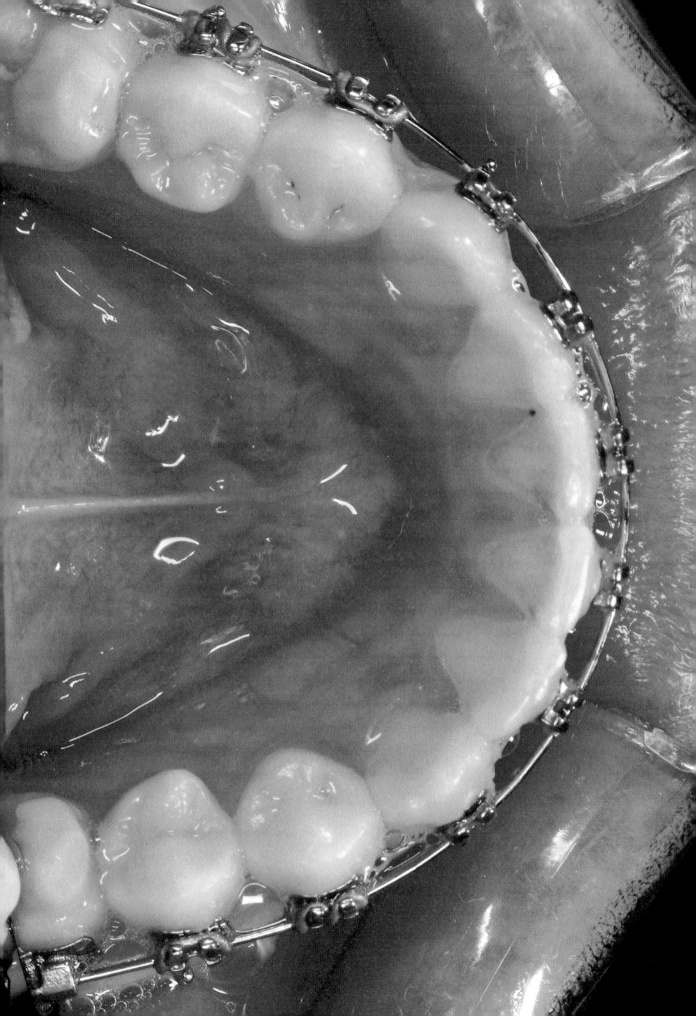

HESTER BIJL, SANDER VAN ZUIJLEN, AEROSPACE ENGINEERS

COMPUTATIONAL ELASTICITY: DYNAMIC INTERACTION BETWEEN FLOWS AND STRUCTURES

1 INTRODUCTION

In many engineering applications, fluid-structure interaction phenomena play a key role in the dynamic stability of a structure (e.g. aircraft, wind-turbines, suspension bridges, etc.). A fast and accurate computation of the dynamic interaction between flow and structure is therefore of the utmost importance. For fluid-structure interaction, often a partitioned approach is chosen to resolve the coupled problem, as it allows reusing existing flow and structure solvers and independent development and optimization of the codes. The drawback of a partitioned approach over a monolithic approach is that the coupling between the flow and structure domain needs additional attention and for strongly coupled physical problems, sub-iterations are required. Sub-iterating increases the computational expense as flow and structure have to be resolved multiple times each time step. Basic sub-iteration techniques include block-Gauss-Seidel iterations, which may suffer instability or fixed under-relaxation methods, which are robust, but at the price of slower convergence. In the literature, several methods can be found for performing sub-iterations in an efficient and robust fashion. One of the most popular methods is the Aitken under-relaxation method, which tunes the under-relaxation parameter to obtain faster convergence. In this paper we apply this model to a two-dimensional strongly coupled laminar and turbulent flow problem and investigate the combination of multilevel acceleration with Aitken under-relaxation.

2 COUPLED PROBLEM

In this paper a fluid-structure interaction problem is addressed which is based on the benchmark problem by Turek. The original problem consists of an incompressible fluid around a circular cylinder with a flexible trailing flap. The coupled problem consists of a fluid domain Ω_f, see figure 14, which is here modelled as a compressible

⑭

the structure and fluid dynamics respectively. In the partitioned approach, the interface boundary between the fluid and the structure domains Γ_I is denoted by two boundaries Γ_{sf} and Γ_{fs} which close the structure and fluid domains so that each domain can be treated separately from the other. The coupling between flow and structure is introduced in the boundary conditions that are imposed on Γ_{sf} and Γ_{fs} and that should yield continuity of displacement (of the interface) and stresses. Since the flap has the highest flexibility in the y-direction and the shear stresses mainly act in the x-direction, we simplify the continuity of stresses to a continuity in pressure so that the conditions at the interface are

$$\mathbf{d}\Gamma_{fs} = \mathbf{d}\Gamma_{sf,} \tag{1}$$

$$\mathbf{p}\Gamma_{sf} = \mathbf{p}\Gamma_{fs,} \tag{2}$$

where d denotes the displacement of the interface boundary and p the pressure. At the moment it is still assumed that the spatial coupling is continuous and the temporal coupling instantaneous. Since the domains have been split and the coupling is performed using boundary conditions, readily available flow and structure solvers can be used to discretize and resolve their own dynamics on their own domains. Therefore, we do not address the specific spatial discretization of the solvers and simply write

$$\frac{\mathrm{d}\mathbf{w}_s}{\mathrm{d}t} + \mathbf{D}_s\left(\mathbf{w}_s, \mathbf{p}\Gamma_{sf}\right) = \mathbf{S}_{s,} \tag{3}$$

$$\frac{\mathrm{d}\mathbf{w}_f}{\mathrm{d}t} + \mathbf{D}_f\left(\mathbf{w}_f, \mathbf{d}\Gamma_{fs}\right) = \mathbf{S}_{f,} \tag{4}$$

wherein $\mathbf{w}_s$ and $\mathbf{w}_f$ are the discrete state vectors for the structure state and fluid state respectively. They contain e.g. the structural displacement or the fluid density. The spatial discretization of the governing equation is simplified by the operator $\mathbf{D}$, which depends both on the state $\mathbf{w}$ and on the fluid-structure interface conditions $\mathbf{p}\Gamma_{sf}$ (the discrete pressure acting on the structure) and $\mathbf{d}\Gamma_{fs}$ (the displacement of the fluid domain boundary). The right hand side may contain terms $\mathbf{S}$ that may arise from boundary conditions on Γ_s and Γ_f. Equations (3) and (4) are in semi-discrete form. We assume that the time integration is performed by the same implicit scheme in both domains. The structure and flow solver programs can then be described as solution techniques that can find solutions $\mathbf{w}_s{}^{n+1}$ and $\mathbf{w}_f{}^{n+1}$ under the boundary conditions $\mathbf{p}\Gamma_{sf}$ and $\mathbf{d}\Gamma_{fs}$ such that they satisfy (or minimize)

$$\mathbf{r}_s\left(\mathbf{w}_s^{n+1}, \mathbf{p}_{\Gamma_{sf}}\right) - \mathbf{s}_s = \mathbf{0,} \tag{5}$$

$$\mathbf{r}_f\left(\mathbf{w}_f^{n+1}, \mathbf{d}_{\Gamma_{fs}}\right) - \mathbf{s}_f = \mathbf{0,} \tag{6}$$

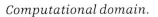

Computational domain.

fluid, and a structure domain Ω_s which is modelled as a linear elastic body. The boundaries of the domain are given by Γ_s, Γ_f, which result in boundary conditions for

wherein $\mathbf{r}$ the residual function (discretized representation of the governing equations), $\mathbf{s}$ a constant source term within the time step that can depend on e.g. previous solutions or boundary conditions, $\mathbf{p}\Gamma$ the discrete pressures in the boundary nodes and $\mathbf{d}\Gamma$ the discrete displacements of the boundary nodes. The subscript s, f denotes that the discrete quantities belong to the structure and fluid domains respectively. A Computational Structure Dynamics (CSD) package is capable of finding a $\mathbf{w}_s^{n+1}$ such that (5) is satisfied for a given pressure load $\mathbf{p}\Gamma_{sf}$. A Computational Fluid Dynamics (CFD) package is able to find a $\mathbf{w}_f^{n+1}$ such that (6) is satisfied for a given boundary displacement $\mathbf{d}\Gamma_{fs}$. A fully implicit (or fully coupled) solution would require

$$\mathbf{r}_s\left(\mathbf{w}_s^{n+1}, \mathbf{p}_{\Gamma_{sf}}^{n+1}\right) - \mathbf{s}_s = \mathbf{0}, \tag{7}$$

$$\mathbf{r}_f\left(\mathbf{w}_f^{n+1}, \mathbf{d}_{\Gamma_{fs}}^{n+1}\right) - \mathbf{s}_f = \mathbf{0}, \tag{8}$$

wherein the superscript $n+1$ denotes the discrete solution at the new time level t_{n+1}. The coupling between (7) and (8) now poses a problem in a partitioned approach as the pressure acting on the structure interface $\mathbf{p}\Gamma_{sf}^{n+1}$ depends on the fluid state $\mathbf{w}_f^{n+1}$ and the displacement of the fluid boundary $\mathbf{d}\Gamma_{fs}^{n+1}$ depends on the structure state $\mathbf{w}_s^{n+1}$. Both the spatial coupling (transferring data from the flow to the structure mesh and vice versa) and the temporal coupling (obtaining an implicitly coupled solution) are addressed in the next sections.

2.1 SPATIAL COUPLING

The coupling between flow and structure takes place at the fluid-structure boundary. In the continuous case, this boundary Γ_I would be identical for both fluid and structure domains; however, at the discrete level, the boundary Γ_{sf} and Γ_{fs} do not have to be matching and gaps or overlaps may occur. A consistent mesh interpolation was preferred over the conservative interpolation method; for consistent interpolation, two separate interpolations are defined for the transfer from fluid to structure mesh and from structure mesh to fluid mesh. The interpolation methods do not have to be the same kind, e.g. a radial basis function interpolation can be used for transfer of displacements from the structure to the flow mesh and a simple nearest neighbour algorithm can be used to transfer pressures from the flow mesh to the structure mesh. For the conservative approach, on the other hand, one can choose an interpolation method for transferring displacements from the structure to the flow, but one has to use the transposed of the interpolation to transfer forces from the flow to the structure. First the coupling of displacements is performed by transferring displacements from the discrete structure boundary to the discrete flow boundary through an interpolation

$$\mathbf{d}_{\Gamma_{fs}} = \mathbf{I}_{fs}\left(\mathbf{d}_{\Gamma_{sf}}\right) \tag{9}$$

The displacements at the structure boundary follow directly from the structure state vector $\mathbf{d}\Gamma_s = g(\mathbf{w}_s)$. When the structure state vector $\mathbf{w}_s$ already contains the displacements of the boundary nodes as part of its degrees-of-freedom, g could simply be represented by a Boolean matrix that extracts only the boundary displacement from the structure state vector.

The second part of the coupling is the transfer of pressure loads from the flow to the structure by an interpolation $\mathbf{I}_{sf}$

$$\mathbf{p}_{\Gamma_{sf}} = \mathbf{I}_{sf}\left(\mathbf{p}_{\Gamma_{fs}}\right) \tag{10}$$

and the pressure at the fluid boundary follows directly from the fluid state vector $\mathbf{p}\Gamma_f = f(\mathbf{w}_f)$. This time the function f may be more complicated as the fluid state $\mathbf{w}_f$ can be defined in cell centers (for a cell-centred finite volume solver), whereas the pressures on the interface may be defined in face centers or vertex locations. In that case an interpolation from cell centred values to boundary values is also taking place in the function f. Additionally, the fluid state vector may only contain the conservative variables and not the primitive variable p, in which case f also computes the pressure from the conservative variables.

2.2 TEMPORAL COUPLING

In partitioned fluid-structure interaction, obtaining the coupled solution described by (7) and (8) would require sub-iterating, e.g. when a sequential algorithm is used

$$\mathbf{r}_s\left(\mathbf{w}_s^i, \tilde{\mathbf{p}}_{\Gamma_{sf}}^i\right) - \mathbf{s}_s = \mathbf{0}, \tag{11}$$

$$\mathbf{r}_f\left(\mathbf{w}_f^i, \mathbf{I}_{fs}\left(\mathbf{g}\left(\mathbf{w}_s^i\right)\right)\right) - \mathbf{s}_f = \mathbf{0}, \tag{12}$$

wherein the superscript i denotes the i-th sub-iteration and $\mathbf{p}\Gamma_{sf}$ is the *estimation* of the fluid pressure acting on the structure for the i-th iteration. The simplest choice for the estimation is

$$\tilde{\mathbf{p}}_{\Gamma_{sf}}^i = \mathbf{I}_{sf}\left(\mathbf{f}\left(\mathbf{w}_f^{i-1}\right)\right) \tag{13}$$

which results in a block-Gauss-Seidel type of iteration, but which is not guaranteed to be stable. To increase robustness under-relaxation can be applied, but generally at the expense of slower convergence rate. In this paper we focus on the widely applied Aitken method, which applies an adaptive under-relaxation to the estimation for the next time step

$$\tilde{\mathbf{p}}_{\Gamma_{sf}}^{i+1} = \tilde{\mathbf{p}}_{\Gamma_{sf}}^i + \theta^{i+1}\left(\mathbf{p}_{\Gamma_{sf}}^i - \tilde{\mathbf{p}}_{\Gamma_{sf}}^i\right) \tag{14}$$

for which the under-relaxation parameter θ^{i+1} is obtained from

$$\theta^{i+1} = \theta^i\left(1 - \frac{(\Delta e^i)^T(e^i)}{(\Delta e^i)^T(\Delta e^i)}\right) \tag{15}$$

with $\mathbf{e}^i = \mathbf{p}\Gamma_{sf} - \mathbf{p}\Gamma_{sf}$ the error between the estimated and the resulting pressure after solving (11) and (12) for iteration i and $\Delta\mathbf{e}^i = \mathbf{e}^i - \mathbf{e}^{i-1}$. For the first under-relaxation step a has to be chosen as $\mathbf{e}^{i-1}$ is not available yet. One can use the last known value and at the very start of the computation any (sufficiently small) value can be taken.

2.3 SUB-ITERATION ALGORITHM

The algorithm that is used in this paper for performing

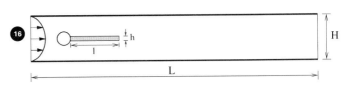

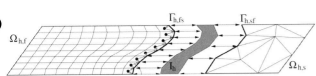

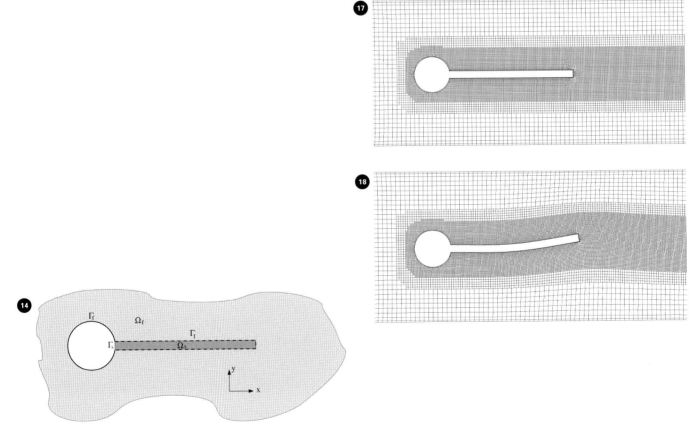

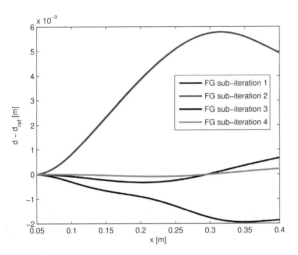

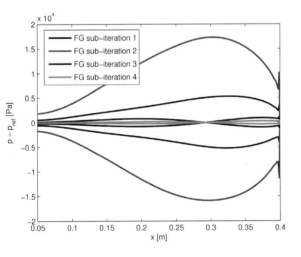

Aitken sub-iterations consists of the following steps and is schematically represented in figure 15:

16

15

Schematic representation of coupling between fluid and structure domain

1. Solve the structure (11) using Aitken for the prediction of the pressure force

$$\mathbf{r}_s\left(\mathbf{w}_s^i, \tilde{\mathbf{p}}_{\Gamma_{sf}}^i\right) - \mathbf{s}_s = 0$$

2. Transfer the displacements of the structure inter face to the fluid interface (9)

$$\mathbf{d}_{\Gamma_{fs}}^i = \mathbf{I}_{fs}\left(\mathbf{g}(\mathbf{w}_s^i)\right)$$

3. Deform the fluid mesh and solve the fluid

$$\mathbf{r}_f\left(\mathbf{w}_f^i, \mathbf{I}_{fs}\left(\mathbf{g}(\mathbf{w}_s^i)\right)\right) - \mathbf{s}_f = 0$$

4. Transfer the pressure of the fluid interface to the structure interface (10)

$$\mathbf{p}_{\Gamma_{sf}}^i = \mathbf{I}_{sf}\left(\mathbf{f}(\mathbf{w}_f^i)\right)$$

17

5. Compute $\mathbf{e}^i$ and when $\|\mathbf{e}^i\|$ larger than some tolerance, continue at 1, otherwise start new time step.

The algorithm converges to a fully coupled solution as the tolerance in step 5 is set to smaller values.

3 RESULTS
The proposed technique is applied to two strongly coupled two-dimensional test problems. The first is a laminar test case and the second a turbulent test case. The flow solver is a unstructured, cell-centred, finite volume solver with a second order central discretization and artificial Jameson dissipation. Coupling of the CFD code to the linear structure code is performed by FLECS, an open source flexible coupling shell that allows the coupling of different codes and performs the interpolation between the interface meshes. The interpolation between flow and structure meshes ($\mathbf{I}_{sf}$ and $\mathbf{I}_{fs}$) is a radial basis function interpolation with a thin-plate-spline function. The structure has its unknowns in the element nodes which are already on the boundary of the elements. Therefore, the function g is simply selecting the boundary displacements from the structure state vector. The fluid unknowns are located in the cell centers and, therefore, the function g performs an interpolation of the cell centered data to boundary vertex data.

3.1 TWO-DIMENSIONAL LAMINAR FLOW PROBLEM
The laminar case consists of a circular cylinder of diameter 0.1m in a channel with height $H = 0.41$m, length $L = 2.5$m, with an elastic flap behind it of length $l = 0.35$m and thickness $h = 0.02$m, see figure 16.

Two-dimensional laminar testcase.

The inflow is a parabolic velocity profile with a mean velocity of 2 and a maximum velocity of 3m/s. The reference Mach number based on the mean flow velocity is set to $M_0 = 0.14$. The Reynolds number is based on the mean velocity and cylinder diameter and is $Re = 200$. The structure is modeled as a linear elastic structure with a density equal to the flow density $\rho = 1000$kg/m^3 and a Young's modulus of $E = 5.6\ 10^6$kg/(m.s^2). Time integration is performed by an implicit, third-order accurate, multistage Runge-Kutta scheme with a time step $\Delta t = 0.01$s. Each implicit Runge-Kutta stage is sub-iterated until $\|\mathbf{e}\| \leq 10^{-2}$. For this test problem the transient to a periodic state is simulated in 500 time steps. The flow solver uses 3 grid levels for the multigrid (MG) solver.

The flow is solved on a computational mesh of 20737 cells figure 17. During the simulation the elastic flap starts to deform and the interior mesh is deformed accordingly using a radial basis function approach, figure 18.

18

(17) Undeformed mesh (18) Deformed mesh Computational fluid mesh.

During the simulations the averaged under-relaxation factor is determined. This value is a measure of how strongly coupled the simulation is. A value of close to 1 means that hardly any under-relaxation is required, whereas a small value of indicates a strong coupling with much under-relaxation. The averaged value of the under-relaxation $\theta = 0.321$, indicating that the problem is strongly coupled and that standard Gauss-Seidel iterations diverge.

The next property that is investigated is the convergence in interface pressure and displacements. The difference in the interface pressure and displacement with respect to the final (fully coupled) solution is shown in figure 19 and 20. In 19 and 20 the results for the Aitken (FG) sub-iterations are shown. It can be seen that at first the error in pressure and displacement increases, as for the first Aitken sub-iteration the under-relaxation parameter $\theta = 0.7$, which is too high for this strongly coupled problem as explained in the previous section.

Differences in interface displacement and pressure
(19) Interface displacement difference
(20) Interface pressure difference

5 CONCLUSIONS

In this paper we investigated the Aitken underrelaxation technique applied to a two-dimensional strongly coupled laminar test problem. The method shows to be capable of solving strongly coupled aeroelastic problems.

THE ESSENCE OF RUNNING

I was a very active child and have been an avid runner for the last 35 years. As a teenager, I became curious about how my body functioned. In particular, I was driven to understand how the muscles, tendons and bones, lungs and heart are integrated to allow a person to run. No wonder then, that I became a scientist specializing in the physiology and biomechanics of running. Elasticity is an essential prerequisite for running. For me to feel fully human as both an athlete and a scientist, I must run. Thus, elasticity is critical for me to be.

Running is defined as a bouncing gait, i.e. a pattern of moving on legs. Bouncing means that the body's centre of mass reaches its lowest point in space, during the middle of the stance phase, when one foot is on the ground. The centre of mass reaches its highest point during the middle of the aerial phase when neither foot is on the ground. At the very beginning of the stance phase, the body has its maximum amount of kinetic energy and a large amount of gravitational potential energy. During the stance phase, the hip, knee and ankle joints all flex and the body centre of mass velocity slows and reaches its nadir (or low point). Thus, the kinetic and gravitational potential energy levels are least at the middle of the stance phase. The flexing of the joints during the first half of the stance phase stretches the tendons, notably the Achilles tendon and the patellar tendon connected to the kneecap. The kinetic and gravitational potential energy is not lost or dissipated; rather, the energy is converted into elastic strain energy and stored briefly in the elastic tendons. Then, during the second half of the stance phase the elastic tendons and ligaments recoil and wonderfully re-convert the elastic energy back into kinetic and gravitational potential energy. That is, the recoiling elastic structures accelerate our bodies forward and push us back up into the air. The elastic structures in our bodies also allow a runner to have a softer landing from each brief aerial phase as gravity pulls us back to the ground.

To appreciate the biomechanics of running, it is helpful to briefly contrast running with the other human gait, walking. In walking, there is always at least one foot on the ground and thus there is no aerial phase. Further, in walking, the body centre of mass moves with an opposite vertical pattern as running; there is no bouncing. In walking, after each foot is placed on the ground, the body centre of mass follows an upward arcing path, reaching its highest point in space at mid-stance. In walking, the impact of the foot with the ground is not effectively cushioned with natural elastic mechanisms. Compared to running, walking is truly 'pedestrian' in the pejorative sense of the word. There have been some very intriguing recent studies which show that walking does involve elastic energy recycling, but the effect is likely much smaller than in running and does not allow a person's body to fly even briefly.[1]

In running, the elastic energy recycling mechanisms are more efficient than any hybrid automobile or ecological recycling process. The ultimate elastic

1 Geyer, H., Seyfarth, A. and Blickhan, R. 'Compliant leg behaviour explains basic dynamics of walking and running Proceedings of the Royal Society B.' 273: 2861–2867, 2006.

2 Dawson, T.J. and Taylor, C.R. 'Energetic cost of locomotion in kangaroos.' Nature. 246: 313–314, 1973.

3 Kram, R. and Dawson, T.J. 'Energetics and biomechanics of locomotion by red kangaroos (Macropus rufus).' Comparative biochemistry and physiology B. 120: 41–49, 1998.

4 Alexander, R. McN. Elastic Mechanisms in Animal Movement. Cambridge Univ. Press, 1988.

5 Roberts, T.J., Marsh, R.L., Weyand, P.G. and Taylor, C.R. 'Muscular force in running turkeys: The economy of minimizing work', Science. 275: 1113–1115, 1997.

6 Kram, R. and Taylor, C.R. 'Energetics of running – a new perspective.' Nature. 346: 265267, 1990.

bouncing animal is the red kangaroo. Back in 1973, two of my mentors, Terry Dawson and Dick Taylor measured the metabolic energy consumption of kangaroos while they hopped on a treadmill.[2] Can you imagine their audacity? The kangaroos wore lightweight masks so that their rates of oxygen consumption could be measured. Prior to this experiment, in all other species of animals, as running speed increases, the rate of oxygen consumption increased. It came as a huge surprise to Dawson and Taylor, and to all other locomotion scientists, when they found that as the kangaroos hopped at faster treadmill speeds, their rates of metabolic energy consumption did not increase! The equivalent would be if a person could run a four-minute mile with the same effort as an easy jog. Dawson and Taylor speculated that the key to the remarkable energetics of kangaroo hopping originates in the elasticity of their magnificently long and thin Achilles tendon.

As a student, I further tested the limits of the kangaroo's exquisite abilities by hopping them uphill on a motorized treadmill. I found in that case they did consume oxygen much more rapidly, presumably because they could not perform the mechanical work to raise their centre of mass vertically using only stored elastic energy.[3] The discovery by Dawson and Taylor stands as one of the major milestones in the science of animal locomotion and many scientists have followed on to learn more about the elastic mechanisms involved.

Kangaroos are extreme in their ability to store and recover elastic energy, but every other mammal and bird that runs uses the same basic process. R. McN. Alexander and his team have measured the mechanical properties of tendons of a wide variety of species for many years. They have found that the tendon is not perfectly elastic, but only about 7% of the energy stored is lost as heat during the recoiling process. Thinner and longer tendons store/return more energy than thick short ones, but they are all made of the same protein, mostly collagen. I would be remiss in an article about elasticity not to mention that Dr. Alexander has thought about and written more clearly about biological elasticity than anyone else. His book, *Elastic Mechanisms in Animal Movement* would be a great place to start learning more.[4]

Tom Roberts and his colleagues devised clever ways measure the forces and length changes in the tendons and muscles of animals while they are running.[5] They have found many situations where the leg muscles are active and generating force but the muscle fibres themselves do not change length very much. Since mechanical work is the product of force and length, muscle fibres that do not change length are doing little or no work. Yet we all know that our muscles demand energy and tire when we run too fast. The reason is that muscles consume metabolic energy whenever they are active and generating force. Muscles do consume more energy when they shorten and perform work but rapidly activating/deactivating a muscle that acts only to generate tension is itself very costly.

Around 1990, Dick Taylor and I discovered a simple quantitative relationship between the rate of metabolic energy consumption and the time of force generation.[6]

The rate of metabolic energy consumption per unit body weight is equal to a 'cost coefficient', c, multiplied by the inverse of the foot-ground contact time (t_c). Remarkably, the cost coefficient is nearly the same 0.2 Joules/Newton in animals ranging in size from small rodents or birds to human horses, and across their entire aerobic speed range. As we run faster, the time of foot-ground contact decreases and our rate of metabolic energy consumption increases with $1/t_c$. It seems that the need to generate muscular force more rapidly at faster running speeds explains why we all get tired when we try to run too fast. But that is only true if there is substantial elastic energy storage and return via elastic mechanisms.

Without elastic mechanisms, running would be far too expensive to sustain because our muscles would have to perform so much mechanical work with each step.

21 You can demonstrate that to yourself with two simple exercises. The first begins by standing in front of a chair as though you had just stood up. Rapidly sit down and stand back up without any pause at the bottom of the movement. Repeat that exercise again and again for a minute. You should pause briefly in the standing position with each repetition. I predict that you will not feel terribly tired. The second exercise begins in the sitting position. Next, stand up and then sit back down rapidly and repeatedly. Pause briefly during each sit and allow your leg muscles to relax. I predict that you will quickly become tired. In both exercises, the same amount of work is performed to lift the body weight. But in the first exercise (stand-sit-stand-pause) you are utilizing the elasticity in your leg muscles to perform the work via recycling. In the second exercise (sit-stand-sit-pause) the tension in the muscles is lost during each cycle and so the elastic energy is dissipated or lost. Thus the muscles must do much more work, which consumes much metabolic energy/oxygen.

I was a curious boy who liked to run around a lot. I liked to stretch my imagination and also my physical abilities. Over the years, continuing to be a curious boy has been crucial to my success as a scientist. Elasticity has also been critical to the discoveries and calculations that I have made as a scientist. Every day that I run stretches my physical abilities and my body responds in adaptation to be more fit. In my 48 years, first as a boy and then as a runner, I estimate that elasticity has cushioned me from a harsh impact with the ground something like 100 million times. During the many hours I have spent running, elasticity has also allowed me to frequently reach the state of being pleasantly tired, rather than becoming exhausted after 100 metres. More metaphorically, running has helped me every day to withstand and rebound from the harsh impacts of adult life. Running depends on elasticity. I depend on running to keep my spirit to be elastic and thus start each day as a curious boy who likes to run.

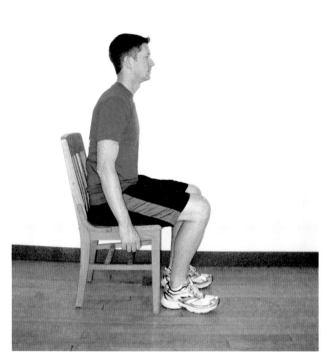

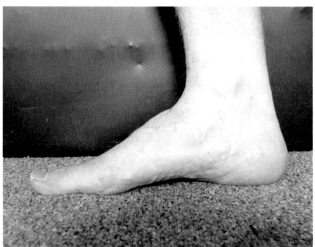

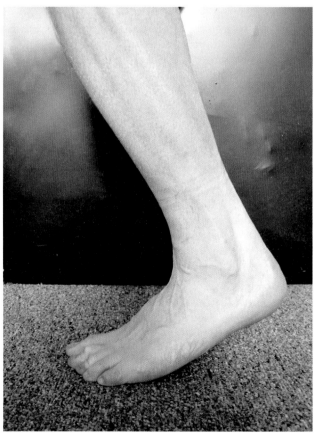

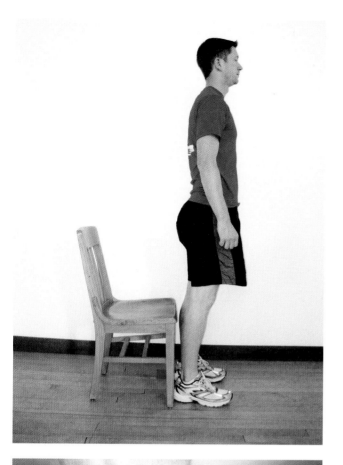

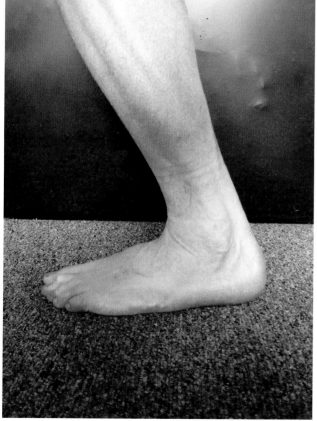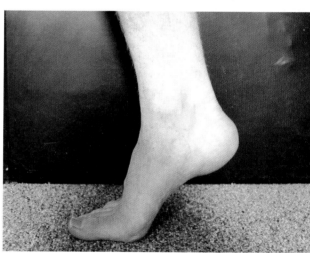

ME AND MY BODY

Complaint: midfoot pain on the left when jumping and pirouetting since yesterday, stiff in the morning, midfoot eversion provocation, also translation MT1,2, sensitivity soft tissue to palpation between MT1,2; squeeze test MTPs neg. MT1,2 mobilisation, provides relief, tape, no jumping or relevéing.

29–1–2002 kick in the face dizziness nystagmus double vision consultation Verstappen is going there today advice not to perform update X–rays were good.
12–2–2002 neck complaints due to slumping posture after dancing ctt traction and c1–2 rot manip and ri c6–7 rot manip th key pointshas to sit in a different position though and do th exerc in future (keep an eye on this)
16–4–2002 pain radiating from the prox. tib.fib. jnt, resulting i numbness lat foot edge Also has lsc complaints. Particularly discomfort when pointéing and after having sat down Provocation test lat cmp. is pos.
Diff diagnosis neuropathy n. peroneus, lat meniscus, radix involvement
Treatment wait and see avoid provocation and lsc extension trng th.
18–4–2002 pain is less but now also pain when walking at fibula level sometimes radiating down to foot LSC ext is painful ALso pain in the ri hip ventrally Manip tlt and ri l3–4 funct impr.
20–4–2002 trng th lsc adapted tlt hip mp is also better it ease!
21–5–2002 complaint the same. Can still be provoked with knee flexion viox is effective but afterwards the same again. Knee mob— viox and medical labelling after England Neurologically there are only sens disturbances continue to retestt strength
19–9–2002 great trauma ri ankle, no fracture, considerable swelling and not possible to put burden on it, pot tendon injury,
12–10 MV. Good recovery lat ankle lig distortion.Mob. UTJ. Also US aplic.
22–10–2002 according to Medhi swelling is decreasing, dorsal flexion and plantar flexion still limited, mnly resistance inversion still provocative, mobilisation UTJ and tape, is overburdening!! is rehearsing 6hpd!!
24–10–2002 discussed overburdening, holding back during lessons and rehearsals no jumping/running less swelling again, mob utj and tape
29–10–2002 says not to have pain, still attends regular fitness training sessions, inversion resistance remains provocative, mobilisation utj and tibiofibular (very stiff)
16–11–2002 is going reasonably well, sometimes shooting pain when medial deep pliéing, no tingling signs, n.tibialis tenderness icw swelling,
19–11–2002 no more pain medial mob.
21–11–2002 is experiencing more pain, very stiff when pliéing during lessons, is becoming a bit frustrated, mob UTJ
26–11–2002 ankle is staying more or the same delayed recovery must strongly improve during the holidays
28–11–2002 is doing better, inversion no longer prov., less stiff when pliéing
3–12–2002 is doing better after treatm with mob to dorsal flex and stretching the hal longus, has the feeling it should snap if this happens it does feel better and more free
5–12–2002 blocked feeling lateral side ankle when walking, mob cuboid to dorsal and MT4,5 dorsal, function improved

7–12–2002 lim remains medial dorsal I thought tib post (strength is good) mob. gives a click but fhl after all in any case mob and stretching the tib/fhl is going better the holidays will do him good must continue to stretch though
12–12–2002 has more pain again, now also with medial point and shooting pain lateral on the calcaneus, palpation of proximal flexor tendon of malleolus prov. these complaints
19–12–2002 experienced painful clicks in the ri shoulder yesterd during lifting, light painful arc, resistances tests neg., prov. Hawkins test, light impingement complaint
10–9–2004 hip is doing better: quadrant technique in abduction, stretch flex hall and mob dorsal capsule, will call Verstappen himself for appointment icw ganglion/cyst? on flexor hall longus
13–9–2004 greatly troubled by burning pains on the location itself, doing a bit better today, mob UTJ, LTJ and stretch fl.hall
17–9–2004 gets echo
20–09 Maintenance.
28–9 swelling is decreasing. in wrist foot frictionand mob
2–10–2004 lot of pain in foot, echo results not yet known.
04–10 foot maintainance and mob. great toe. Also mob. wrist.
5– image of fasc plantaris or nodular or torn muscle after all? can't do anything with this> amc
12–11–2004 thoracic limitations: manip T 5/6, T8/9, TLT, ri costa 8, ri/ri l3/4, and ri ventral SI
16–11 Manip th 6 8 10 + 2nd rib ctc ri ventral si
19–11 after many lifts in short time ri rib 4–5 manip tltand th 8 manip.ri and ri ventral si.
20–11 is doing better ctc beast mp rib 7–8 on the right does not reach mob and massage impr function nxt time mob again?
23–11 lback is a bit better mob sien costae th
26–11–2004 a little stiffer again: manip 6/8 and mob costae in prone pos
2–12 ri foot problem lateral blockage and cramping calf mob cub. and calf massage.
3–12–2004 foot better, calf again, also mid back again mp T5/6, T7/8, mob costovertebral burdened 3d heteronome and TLt 3d ext
6–12 Now complaint CSC partic. in CTC region flexion and left rotation are blocked the most.M. T3/7 +/+
10–01–05 Acuute block. M T9 +/+
11–1 manip. th 56–7 act mob.
13–1 much btter now pelvis to ventral and gap tlt
2–2–2005 mob ri SI and mob ri L4/5
4–2–2005 CSC blockage: nelson and mob CTC in flexion
11–2–2005 ri shoulder impingement, manip ri costa 2, abduction mob and 1st ri rib
21–2–2005 rec pelvic complaint
22–05 mob si and l5–s1
24–2 doing a little better
12–4 tlt mp and ri ventral ri.
13–04–2005 mob SI, stretch illiopsoas and rectus fem/tensor
18–04 Complaint CTC acute since Friday. Block. T2 mob. and Manip.
19–4 th 4 manip and ctc rot. manip.
20–4–2005 does not want any mp for now in CSC, mob CTC 3d flexion, mob left 1st rib, mp 2nd rib
21–4 nelson mp 2nd rib and th 4 are increasingly better only morning stiffness

persists
19–5 has lifted a lot stiffness and pain ri lsc. Mob ri ventral si si and ap mob. l5 l4 unilaterally and finally flexion mob. l5/s1 funct. improv. will do fewer lifts for a while
20–5–2005 ri SI, stretching
25–5–2005 slight improvement,
6–6–2005 mob ri SI with exorot hip, hold/relax technique ri L4/5, mp TLT, mob plantar flexion UTJ and mob LTJ
19–09 Left hamstring complaint partic. related to hip ca[psule irritation.Musc techn and mob. hip in all directions partic. also abd.
20–9 l5–s1 mob in side position mp l2 and th12 gap friction local funct. improv.
21–9– mob left SI ventrally and locally for a moment
26–09 Block. high ri CSC now mob. 3D hetero flexion with ok result.Also M. Co2 ri T3/5 +/+,mob. is improving well.Possible M. high left CSC if

complaint stagnates.

27–9 persisting pain after mp chris
30–9–2005 maintenance ri ankle: plantar flexion limited
4–10 high th blockage more paiinful aftermassage yesterday th4 ctt and c7 rotmp f. improv.

second time funct. improv ossc. high th.

5–10–2005 is a bit better, now more troubled by blockage high cervical and 3d flexion ri: mp c2/3 and c0/1
6–10 th 4 c3–4 and
7–10 neck hsa improved somewhat, mp ri costa 2/3
14–10 plantar flexion limit and light irritation of peronei
18–10
21–10 blockage csc mp ri C3/4
28–10 acute blockage TCT: gap mp tlt and t7/8,
11–11 recurrent ri groin complaint, strain adductor region,
6–1–2006 acute low back pains due to bad landing after big leap, made kind of 3d hyper extension movement, mp TLT and mob LSC: function improv
7–1 was improved sleeping was very painful TH: stretch iliop soaos mp key points flexion mob flos techn and in supine position stretch m.extensor. Will now particularly exercise flexion and med. advice if no improv by Monday sleeping in side position no X–rays yet nocturnal pain shouldbe gone/decreased by Monday
09–01 Has attended class now and some cramping again. Especially good effect ri 3D ext.
13–1–2006 acute blockage ri costa 3/4, mp costa 3/4 and mp t3/4
18–1–06 Hamstring attachment tightness and pain, worst with high retire, treatment with soft tissue realease and active tissue mobilisation
20–1 hold-relax hamstring and locally
23–01 Acuut block T4/6 since last weekend. Mob. Musc tec hn. M. T4/7 +/+.
24–1 stomach intestinal complaints diarrhea has improved gave ors strong th improv mp th 4 mob 5th rib.

second complaint around the tlt by lifting TH: gap mp l3 l1 th11 and funct. improv. rotaions improv by stretching iliop soas and tlt 3d ext. burdened mob.

tomorrow wanda check

25–1 mp TLT, mp T7/8 and T4/5, mp ri costa 3, mob TLT burdened rotation and CTT burdened extension
28–1–06 Upper hamstring / obturator externus pulling and stiffness. Release gamelli, obturators, fascial release hamstrings, release Add attach
3–3:
14–4–2006 light sprain in ri groin yesterday, mob pelvis and hip
27–4 mob, le SI towards rear
29–4 tlt gap and ri l5–s1.
5–5–2006 light dorsal impingement mob utj and tibiofibul, overburd quadriceps bsds
2–6–2006 neck blockage CTT region, stiffness bsds but no pain
9–6 groin complaint pelvic treatm.
14–08 Maintenance back Mob SIJ both hips M. T4/8/11 +/+ mob is increasing.
23–8–2006 mob le UTJ in plantar flexion
3–10–2006 mp C6/7, mp ctt, mp T4/5
6–102006 ankle acting up again: plantar flexion limited, csc still not totally in order
7–10 rep treatm
22–10 lsc complaint tlt mp and key points therapy for neck
30–11 dorsal knee pain improv now more pain ventrally in le ankle tender tendon tape a strip to support medial vault stretch and mob ankle
17–11
6–1–2007 mob proximal and distal, lateral meniscus?
12–1–2007 would like echo icw persisting dorsal knee pain, mob UTJ, tibiofibular
30–1 mri dviers peter
1–2 persisting tension l5–s1 all mob. and ap and m,assaging techn provide temporary relief nsaid does too much, especially lifting.
2–2–2007 is not going well: burning feeling, can hardly stand on 1 leg, local pain around SIPS, SI provocation tests 3 out of 5 tests recognisable pain, traction mp hip, mob dorsal ilium: no work today
6–2 a little better every day treatment rep and explanation again appears more selective has different med. Will be going for a walk and a swim see him again shortly Thursday decision on any performance
7–2–2007 still some stfifness
8–2 si loose especially 3 d ext left L 3
10–2 is going well did part of the lesson marking on stage. Continue treatment now for tlt and left costae 7–8–9 mp (si dorsal is going well) . cont. medication.
12–02 Good recovery but still tenderness le SIJ region. Partic. still stiffness le lat flexion. Partic. ri post 3 D ext mob TLT gets best results. Must build up slowly.
13–2 felt good but yesterday after working some pain in his back at the l2 level mp/mob tlt l2–3 stretch techn hold relax and mob/mp hip
16–2 mp tlt and mob si also on the right and l5–s1 also on the right
20–2 continues to compain about burning feeling l5–S1 at the peroneus level, IMO it originates from the lsc, feels it especially during extension. mp rib 10 up to and including 12, mp l1, flexion mob l5
2–3–2007 acute lower back pain, seems now l3/4 disc related; slump positive, centralisation during extension, extension exercises and mobilisation in extension provides relief
3–3–2007 is doing a little better, mac kenizie works best, stabilising exercises
05–03 Continuation mob ventral L3/4. Expedient and fast recovery.M. t4/6/9 +/+.
7–3–2007 got too much pain on the bus, could not dance, mac kensie and 3d tlt burdened, stretch iliopsoas
9–3–2007 some progress, discussed exercises again\
10–3 started a lift once with alex on his back from 90 flexion lsc, explosuve without warming up…
12–03 Must not want too much and too fast has to perform with Alex nxt week!
13–3 is going much better recognises the importance of stab exercise therapy is selective mob ri si upon reqest and key points exerc therapy for the multif.
14–3–2007 is going well also after stopping medication
20–4–2007 rec sensitivity around SIPS! mob ri ventral ilium
2–5–2007 shortlasting mob ri ventral ilium
16–08 again experiencing problems with le FHL and also Hallux as a consequence of this. Suggested to perhaps have another check soon.

will discuss this with Bert and Wanda also.

17–8–2007 massage is helpful
20–8–2007 Ledderhose syndrome: type of Dupuytren in the sole of the foot; removal can be considered but often lumps go away by themselves
29–9 acute complaintri hip hip flexion + and endo mp funct. improv hip and mp l2–3 and tlt and mob ventral si
10–10 M. T9/11 +/+
22–10 M. t7/10 +/+
16–11–2007 acute neck blockage on the right high cervical, no headache: mp left C2/3, mob right c1/2 high cervical mobilisations: more pain: prescribed cataflam
7–2 blockage feeling low cerv mp ctt t4 th 8
03–03 Mob ri SIJ partic good effect ventral Ilium.
5–3–2008 sprained left index finger yesterday: evidence of light hematoma palmary especially abduction is painful, swelling, wait out inflammation phase
23–5–2008 blockage feeling TLT: mp TLT and T8/9, mob TLT in side position and burdened
4–6–2008 inflammation nail wall great toe: monitor bacterial infection: will use fucidine now, so far no constant pain but mainly pressure pain, possibly medication tomorrow?
23–8–2008 some trouble with lef knee. got this yesterday, unnown reason. better this morning, but after half an hour in class a little more pain again. some complaints m popl and prox tib/fib. mob ventr/lat and mus ctechnique popl. improv flexion, flexion/exo passive and endo rot active of tib. also gave self taechniques for muscle and mob
25–08 Still some feeling of instability no provocable heavy disorders.Mob Flexion ext.
26–8–2008 is doing a bit better. thoracic bit stiff and tlt too. mp th 8,10 and tlt
22–09 some light irritation dorsal capsule knee and ri pes anserinus.Mob knee.
08–12 Blockage complaint ri SIJ mob. and T3/6 M. +/+
9–12 shock wave for musc and mp coate 4 to

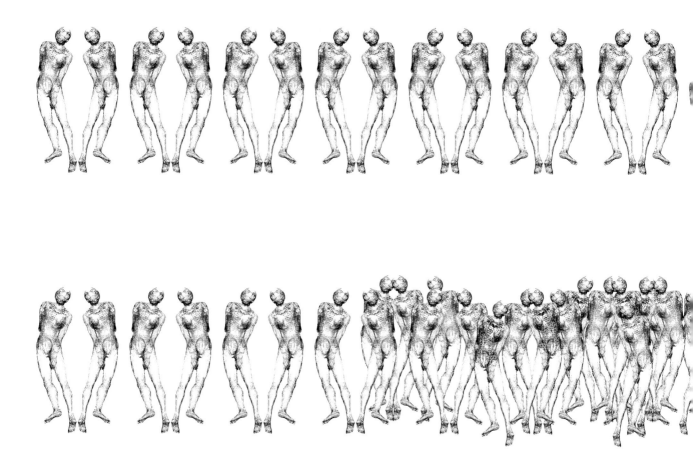

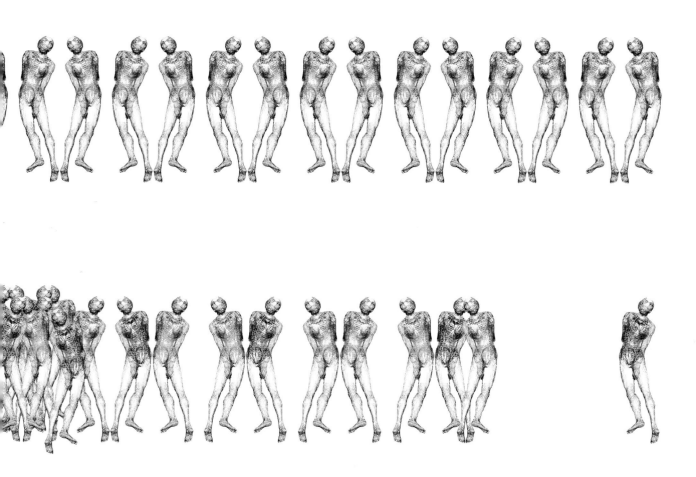

I can't say the word *relationship*

SHFAQAT ABBAS KHAN, GEOPHYSICIST

A TOUGH CRUST

The Earth is an elastic material or body and is deformed by a variety of sources. Elastic deformations caused by gravitational attraction from the Sun, the Moon and other planets are known as direct elastic deformation. However, the Sun and the Moon also cause harmonic circulations in the ocean, which give an indirect contribution to elastic deformations at the surface of the Earth. Elastic deformations generated by the oceans are known as 'ocean loading'. Similarly to ocean loading, atmosphere pressure loading caused by ice sheets, glaciers etc. can also contribute to elastic deformations, measurable at the Earth's surface by various geodetic techniques such as GPS (Global Positioning System) and VLBI (Very Long Baseline Interferometry).

Modelling the elastic loading deformations requires a representation, in time and space, for the surface load (e.g. ocean, ice and atmosphere) as well as one for the Earth's structure. W. E. Farrell first published the standard method for determination of the elastic loading effect in 1972.

To compute the elastic loading effect from, for instance, the oceans or the atmosphere, a point mass dm is considered, located at a distance α, from the observation point P.

mass is concentrated at a single point. The elastic loading effect is given by,

$$dL(\alpha) = \zeta(\alpha)\,dm$$

The elastic loading effect, dL, represents, for example, gravitational variations, three dimensional surface displacements, changes in the gravitational potential or tilt of the Earth's surface. ζ is the elastic Green's function, describing the effect of a delta-function excitation. The total elastic loading effect or the elastic loading effect from all of the point masses is obtained by integrating over all mass elements,

$$L = \int \zeta(\alpha)\,dm$$

This is also called the convolution integral, since the elastic Green's function is convolved with the mass load distribution.

The rapid unloading of ice from the southeastern sector of the Greenland ice sheet between 2001 and 2006 caused an elastic uplift of ~35 mm at a GPS site in Kulusuk. Most of the uplift results from ice dynamic-induced volume losses on two nearby outlet glaciers for example.

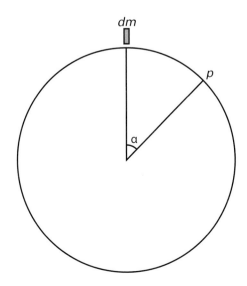

The point load has the unit of 1 kg and the

Reference
Farrell, W. (1972), 'Deformation of the Earth by surface loads', Rev. Geophys and Space Phys, 10, 761–797.

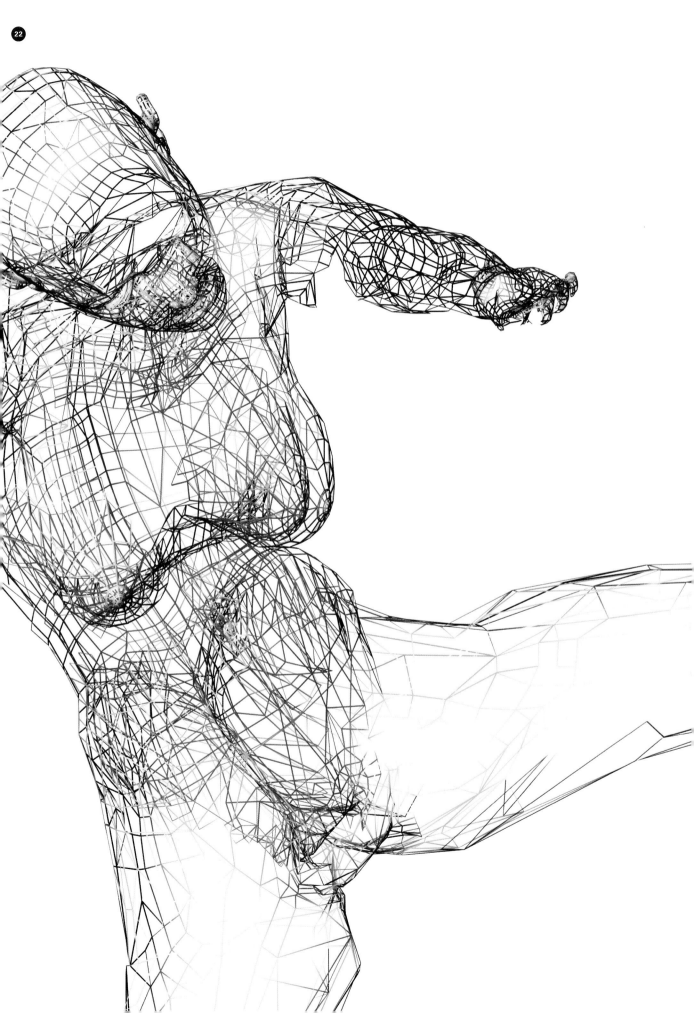

CATHERINE MALABOU, PHILOSOPHER

THE ELASTIC AND THE PLASTIC IN THE PLEASURE PRINCIPLE

In *Beyond the Pleasure Principle* Sigmund Freud presents the concepts of *plasticity* and *elasticity* as synonyms.[1] In reality, plasticity and elasticity designate two different and even opposed relationships between matter and form. A plastic material, once shaped, cannot go back to its previous state. An elastic material on the contrary is able to return to its initial form after undergoing a deformation. The unperceived contradiction between plasticity and elasticity in the text is the very expression of Freud's failure to ascertain the existence of any 'beyond the pleasure principle'.

Plasticity, for Freud, characterizes the indestructibility of our earliest psychic formations. This idea appears very clearly in *Thoughts for the Times on War and Death*

> *Every earlier stage persists alongside the later stage which has arisen from it; here succession also involves co-existence, although it is to the same materials that the whole series of transformations has applied. The earlier mental stage may not have manifested itself for years, but none the less it is so far present that it may at any time again become the mode of expression of the forces in the mind, and indeed the only one, as though all later developments had been annulled or undone. This extraordinary plasticity of mental developments is not unrestricted as regards directions; it may be described as a special capacity for involution—regression—since it may well happen that a later and higher state of development, once abandoned, cannot be reached again. But the primitive stages can always be re-established; the primitive mind is, in the fullest meaning of the word, imperishable.[3]*

The 'extraordinary plasticity' of mental development is thus linked with the permanence of form. Once formed, psychic matter cannot go back to its previous state. We must remember that 'plasticity' generally describes being at once capable of receiving and of giving form. The psyche is plastic to the extent that it can receive the imprint and impose this earlier form upon most recent developments. But we also know that plasticity equally means the power to annihilate form. Plasticity may be used to describe the crystallization of form as well as the destruction of all form (as suggested by the term 'plastic explosive' in bomb-making).

This destructive meaning of plasticity is also present in Freud's characterization of psychic life. Paradoxically, the permanence of form, the impossibility to forget, appears to be a specific means of destruction of this same form. If it is true that a conservative instinct in the psyche exists which tends to restore an earlier state of things, i.e. the inorganic passivity of matter before it came to life, then the status of the plasticity of psychic life is properly undecidable. The impossibility of erasure or disappearance in mental life expresses equally the liveliness of the trace as well as the inertia proper to the death drive. That is why this liveness is also the mask of mental disease.

What are called mental diseases inevitably produces

an impression in the layman that intellectual and mental life have been destroyed. In reality, the destruction only applies to later acquisitions and developments. The essence of mental disease lies in a return to earlier states of affective life and functioning.

The impossibility of oblivion coincides with the inability to change, with the tendency to restore an earlier state of things, and with the deadly mechanism of the compulsion to repeat. We remember this passage from *Beyond the Pleasure Principle* in which Freud declares:

> *The elementary living entity would from its very beginning have had no wish to change; if conditions remained the same, it would do no more than constantly repeat the same course of life. (...) Every modification which is thus imposed upon the course of the organism's life is accepted by the conservative organic instinct and stored up for further repetition. Those instincts are therefore bound to give a deceptive appearance of being forces tending towards change and progress, whilst in fact they are merely seeking to reach an ancient goal by paths alike old and new.[4]*

In *Civilization and its Discontents* (1930), Freud shows that all possible comparisons between the psyche and other types of development are faulty.[5] The plasticity of mental life is first compared with the past of the city of Rome. "Let us, by flight of imagination", says Freud, "suppose that Rome is not a human habitation but a psychical entity with a similarly long and copious past—an entity, that is to say, in which nothing that has come one into existence will have passed away and all the earlier phases of development continue to exist alongside the latest one…" But this comparison is not satisfactory. "There is clearly no point in spinning our fantasy further," Freud goes on,

> *for it leads to things that are unimaginable and even absurd. If we want to represent historical sequence in spatial terms we can only do it by juxtaposition in space: the same space cannot have two different contents. Our attempt seems to be an idle game.[6]*

The plasticity of mental life cannot be represented in 'spatial terms'. The same thing occurs if we compare this plasticity with "the body of an animal or a human being." But here, too, says Freud, we find the same thing:

> *The earlier phases of development are in no sense still preserved; they have been absorbed into the later phases for which they have supplied the material. The embryo cannot be discovered in the adult. [...] In the marrow-bones of the grown man I can, it is true, trace the outline of the child's bone, but it itself has disappeared, having lengthened and thickened until it has attained its definitive form. The fact remains that only in the mind is such a preservation of all the earlier stages alongside of the final form possible, and that we are not in a position to represent this phenomenon in pictorial terms. (pp19–20)*

Nor can the plasticity of mental life be represented in 'pictorial terms'. Organic life strangely suffers from the same defect as architecture. Space is the privileged metaphor for its developments. On the contrary, the plasticity of mental life implies an unpicturable state of things in which emergence and preservation, life and inertia, vitality and passivity coincide in time.

Is there a way to set up a proper representative model for this temporality? If pictorial representation is not satisfactory, can we think of another kind of representation?

In *Beyond the Pleasure Principle*, Freud invokes Ewald Hering's theory:

> *According to Hering's theory, two kinds of processes are constantly at work in living substance, operating in contrary directions, one constructive or assimilatory and the other destructive or dissimilar. [...] We venture to recognize in these two directions taken by the vital processes the activity of our two instinctual impulses, the life instincts and the death instincts. (p49)*

Eros, or the life drive, creates forms. The death drive destroys them. Life drives and death drives are two plastic tendencies that coincide in time. But Freud does not succeed in bringing to light the very form of this temporal and material coincidence. He fails because he is led insidiously to dissociate this simultaneity. At the very moment when he defines the plasticity of mental life as a coexistence of life and death, as an undecidable state between them, he introduces a distinction between plasticity and elasticity, which breaches this undecidability or this coexistence.

If we read *Beyond the Pleasure Principle* closely, we discover that only the life drives are eventually said to be plastic. The death drives are said to be 'elastic'. The destructive tendency, the compulsion to repeat, the restoration of an earlier state of things are eventually driven out of the field of plasticity.

It is noticeable that Freud never uses the words 'plastic' or 'plasticity' to characterize the work of the death drive. In *Beyond the Pleasure Principle*, the death drive appears as 'a kind of organic elasticity, or, to put it in another way, the expression of inertia inherent in organic life' (p36). The problem is that elasticity is precisely the term which Freud makes use of when he characterizes the regulation of pleasure. Elasticity is the name of the tendency to maintain the quantity of excitation at its lowest level. In this sense, elasticity is another name for pleasure, or for homeostasis, i.e. for the pleasure principle. If the compulsion to repeat is able to resist the pleasure principle, it is to the extent that it is not elastic, that it forms a kind of unpleasure that pleasure cannot erase. But instead of a fascinating face-to-face between creative plasticity and destructive plasticity within the compulsion to repeat, we find a disappointing contrast between plasticity and elasticity. Life creates form, death is a formless return to matter. Death is a levelling of all forms, just like pleasure.

Freud states however that the profound meaning of the death drive is the immanence of death in life. Death is not, or not only, an external threat, but it works within life. It means that life forms its own destruction. "The organism," says Freud, "only wishes to die in its own fashion" (p39). The organism fashions or forms its own death. There may be an elasticity of inorganic matter, but it is attained only as the result of a formative process, which is of repetition or the work of the

death drive. But Freud does not succeed in characterizing this formative process, or this fashioning. He never gives an example of such a negative fashioning. Destructive plasticity once again is reduced to elasticity. There is eventually no plastic work of the death drive. No *forms of destruction*.

The impossibility to characterize the form of the death drive constitutes the main objection against its existence. Freud is well aware of that when he writes:

> *The difficulty remains that psychoanalysis has not enabled us hitherto to point to any instincts [or drives] other than the libidinal ones.*

For the moment, we can only prove the existence of erotic drives, that is of life drives, which do not exceed the realm of the pleasure principle. He tries to find what he calls an 'example', that is to say a form, of a death instinct in sadism.

> *From the very first, he says, we recognized the presence of a sadistic component in the sexual instinct. As we know, it can make itself independent and can, in the form of perversion, dominate an individual's entire sexual activity. (pp53–54)*

The form of the sadistic instinct when it 'separates off' the life drives or when it 'has undergone no mitigation or intermixture' may be considered as a possible form of the death drive. "If such an assumption as this is permissible, then we have met the demand that we should produce an example of a death instinct."

However, Freud is clearly not satisfied with this 'example'. Sadism and masochism are still derived from love and proceed from the transformation of love into hatred. In this sense, they still belong to the pleasure principle and express "the familiar ambivalence of love and hate in erotic life" (p54). Sadism and masochism ultimately are and can only be forms of pleasure.

Because he introduces a non-plastic element in his definition of the plasticity of mental life—i.e. elasticity—Freud ruins the possibility of thinking what he precisely wishes to think, the *plastic coincidence between creation and destruction of form*. Either the death drive borrows its plasticity to Eros, and it appears paradoxical as a manifestation of pleasure, as proved by the examples of sadism and masochism, or the death drive functions as an elastic, and it still appears as a manifestation of pleasure. Because Freud deprives the death drive of any autonomous and specific plastic power, he deprives it also of its capacity to resist the pleasure principle. If we are not able to prove that the destruction of form has a form and is a form, if form itself is always on the side of Eros and of pleasure, then it is impossible to prove that there is anything beyond the pleasure principle.

1   *The Standard Edition of the Complete Psychological Works of Sigmund Freud*, James Strachey, 24 vols., London: Hogarth, 1953–74 [SE], SE 18: 1–64
2   Ibid, SE 14: 273–300
3   Ibid, *Vol XIV*, pp285–6)
4   *Beyond the Pleasure Principle*, Sigmund Freud, 1920 S.E., XVIII, p38
5   SE 21, 10–72
6   Ibid 18–19

Play

AN INVITATION TO CONTRIBUTE

Date: 22 February 2010 03:16:33 GMT+01:00
Subject: woo ingi

Dear Mr/Mrs.Astrid van Baalen

I have attached profile of Mr.Kwon, please check it.

Also, I have some questions I would like to ask to you.
First, when are you planning to publish your publication,
Second, are you planning to request Mr. Kwon's performance and if you are, I would like to have his performance when your publication is published.
Actually, we are ready to do the performance and tightrope-walking now.

I am looking forward to hearing from you.
Thank you
Yours sincerely
Woo, Ingi

Datum: 29 mei 2009 18:29:20 GMT+02:00
Onderwerp: Antw.: korea woo inki

Dear Woo Ingi,

We are very sorry to hear of Kwon Won Tae's accident and are wondering how his recovery is going.
We have seen some very impressive photographs of Kwon Won Tae performing and wonder what happened. Did he fall during a performance?

You mention in your email that tightrope walking is a Korean tradition and we wonder if you could tell us a bit more about this? Is there a history of tightrope walking in Korea? Did Kwon Won Tae start at a very young age? Is it a tradition that is passed from father to son? We would love to learn more.

Please pass on our sympathy and best wishes to Kwon Won Tae.

Yours sincerely,
Astrid van Baalen

PS. Would you like to receive our previous publication, *Findings on Ice*, in the post?

Datum: 11 mei 2009 03:10:35 GMT+02:00
Onderwerp: wooingi

Dear
Thank you very much for your email and I am sorry to make you worry about Mr.Kwon WonTae.
I am afraid Mr. Kwon would not be able to contribute to your publication in 2 months.
However, do not hesitate to contact us, please. I would like to keep in touch with you.

I wish you luck.
Thank you.
Yours sincerely
Woo, Ingi

Datum: 1 mei 2009 15:22:41 GMT+02:00
Onderwerp: Antw.: korea woo inki

Dear Woo Ingi,

Thank you very much for your reply. We are so sorry to hear about the accident! Please pass on our best wishes and sincere hope for a quick recuperation to Kwon Won Tae.

Can we also consider if a contribution in two months would be possible?

With kind regards
Hester Aardse

Datum: 30 april 2009 06:27:50 GMT+02:00
Onderwerp: korea woo inki

안녕하십니까?
저는 권원태 선생님의 메니저 우인기 입니다..
이렇게 늦게 메일을 드려서 너무나 죄송합니다.
다름이아니오라 얼마전에 공연중에 사고가 발생하여 몸이 불편한 상태 입니다.
너무 걱정할 정도는 아니지만 약2개월후면 모든 일과 공연이 정상적으로 가능합니다.
그리고 이번 행사에 참여하지못해 무척 유감이지만 어쩔수 없었습니다.
또한 저희는 언제나 마음이 열려있고 좋은 의도의 공연 이라면 언제든지 서로 의논 할수 있다고 봅니다.
추후 세계적인 명인들과 멋진 공연을 함께하고도 싶습니다.
가장 중요한것은 평생 수십년간 해온 권원태 선생님의 열정과 좋은 작품으로 한국의 전통을 알릴수있는 메카가 되도록 여러분 단체와 협력하고 싶습니다.
항상 저희팀에게 관심을 가져주신 여러분께 다시한번 감사드리며
자주 연락 하도록 하겠습니다.
귀하의 단체에 항상 행운이 가득하길 바랍니다.
안녕히 계세요!

Dear Mr/Ms Hester Aardse and Astrid van Baalen

First, thank you very much for your invitation.
I, Woo Ingi, am writing you as a manager of Kwon Won Tae and I appologize for late replying.
Kwon Won Tae has been in hospital due to unexpected accident during his performance. But he would be able to do his performance and his work about after 2 months his doctor said.

It is very regretable that we could not participate to your publication *Findings on elasticity* this time.
However, we always would like to discuss about good performances or publications with you if they are designed with good intention. Furthermore, we would love to have performances with other renowned and emerging artists if there are any opportunities to do

that in the future.

We assume that contributing and working with you, would be a great opportunity for Korean traditions to the world with Kwon Won Tae's great work and performances.

I would like to say thank you again for your attention to us and keep in contact with you often, please.

Yours sincerely
Woo Ingi
manager of Kwon Won Tae
미안 늦어서요..>< 수고해요!!

Datum: 24 april 2009 12:36:43 GMT+02:00
Onderwerp: REMINDER findings on elasticity: an invitation to contribute

Dear Kwon Won Tae,

Recently we send you the invitation to contribute to the international arts and science publication *Findings on elasticity* (please see below).

We can very well imagine this slipped your attention, but we are of course very eager to know your reaction! Please don't hesitate to contact us if you have any questions, we'd be more than happy to answer them.

Sincerely yours,
Hester Aardse

Begin doorgestuurd bericht:
Datum: 17 april 2009 14:03:15 GMT+02:00
Onderwerp: findings on elasticity: an invitation to contribute

Dear Kwon Won Tae,

We are writing you as founders of the Pars foundation. Pars is an independent initiative that views the arts and sciences as essentially creative processes borne out of sheer curiosity. We collect the thought notations of those artists and scientists who shape the way we perceive the world. These are bound in an annual publication, which has a different theme each year and is published by Lars Müller Publishers. Our aim is to create an atlas of creative thinking at the beginning of the 21st century. The first book was *Findings on ice.*

The theme for our next publication is **elasticity**, in the broadest sense of the word. Elasticity is the sound of a piano string being struck hundreds of times without going out of tune, but also the movement of economic markets, the suppleness of a dancer's limbs or a simple device for fixing the hinge on the door.
Pars would very much like to invite you to contribute to *Findings on Elasticity*.
We would be honoured to place your contribution alongside other renowned and emerging artists

and scientists who shape the way we look at the world today.
Please find the list of all those contributing below. Please note the form in which you choose to contribute is completely free. We are not necessarily looking for an already finished piece or something polished; rather we are also looking for the notes in the margin, sketches, images and jottings that lead up to a finished piece: the thought-notations.
There is only one catch: your contribution should reflect the language of your discipline and relate to elasticity.

We want to offer readers a view of what goes on in the cerebral stove of creative thinkers and how all these different disciplinary languages will interact when placed side by side. We strive for a visually and textually attractive publication whilst respecting the integrity of each contribution.

We hope that we have aroused your curiosity. Please let us know if you are interested, or if you have any questions.

Yours sincerely,
Hester Aardse and Astrid van Baalen
Pars Foundation

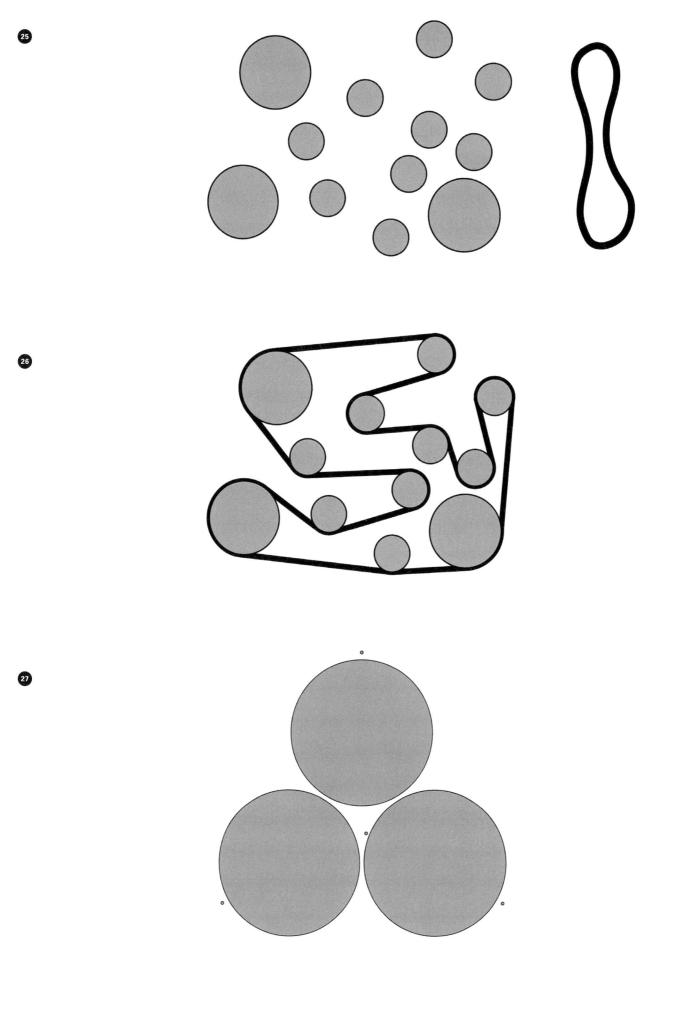

CONVEYER-BELT ALPHABET

ERIK DEMAINE, MARTIN DEMAINE, BELÉN PALOP,
MATHEMATICIANS, COMPUTER SCIENTISTS

CONVEYER-BELT ALPHABET

Mathematics is often pursued purely for mathematical outcomes: theorems, proofs, open problems, conjectures. The open problems beg to be answered, and the quest to solve them is tantalizing and difficult—sometimes provably impossible. When successful, solving a problem is extremely rewarding, because mathematical proof offers (a small part of) the ultimate truth. But the pursuit toward this truth is equally interesting, and the field becomes richer if we allow research on mathematical problems to produce all sorts of outcomes, from art to puzzles to design. By exploring these connections between mathematics and diverse fields outside science, we see a new side to the original problems, leading to inspiration and, hopefully, solution.

The first two authors have found this open-ended approach to be both productive and enjoyable. Some examples are mathematical results in hinged dissection [DDEFF05, DDLS05] that later inspired a new mathematical font design [DD03] and an interactive sculpture [DDP06]; study of pleated origami that led to algorithmic sculpture [DDL99] and just recently culminated in a mathematical surprise that the objects we worked with do not in fact exist as we thought them to [DDHPT08]; study of curved-crease origami that led to sculpture at MoMA [DD08a] and architecturally relevant designs [KDD08]; pursuit of open problems in computational origami that remain unsolved but have led to a series of puzzle designs [DD08b].

Here we describe one such open-ended exploration, on a mathematical problem of wrapping an elastic loop around given wheels, and the mathematics, font design, and puzzles that resulted.

CONVEYER-BELT PROBLEM

Suppose we are given several disjoint disks pinned at their centers in the two-dimensional plane, and a closed elastic band, as in figure 25.

25 The input: disks and an elastic band.

We think of the disk as a wheel that can spin freely around its center (but cannot otherwise move), and the band as a conveyer belt or rubber band, modeled as a stretchable closed loop in the plane that, at rest, tries to contract its length. Now we are asked to wrap the band around the disks in such a way that (1) the band touches every disk and (2) the band is taut, unable to contract in length given the disk obstacles. (see figure 26).

26 The goal: a valid wrapping of the elastic band around the disks.

Equivalently, we want the band to wrap the disks so that rolling the band like a conveyer belt simultaneously turns all of the disks like wheels. More specifically, if the band rolls clockwise in the plane, then disks interior to the belt will rotate clockwise, while disks exterior to the belt will rotate counterclockwise.

The central mathematical question here is 'what arrangements of disks have this kind of proper band wrapping?' Are there simple characterizations of when it is possible, or an efficient algorithm to tell whether given disks have a proper band wrapping? This seemingly simple geometrical problem has been posed by computational geometer Manuel Abellanas at several workshops, the earliest being the 1st Taller de Geometría Computacional in Cercedilla, Spain (2001), where the third author learned of the problem, and the most recent being the Workshop on Computational Geometry in Girona, Spain (2006), where the first author learned of the problem. The problem also recently appeared in print [Abe08]. By now dozens if not hundreds of researchers know of the problem, yet still surprisingly little is known.

The only result so far is that a proper band wrapping does not always exist. Javier Tejel and Alfredo García found the seven-disk example in figure 27. Note that the disks have vastly different sizes.

27 Seven disks of different sizes with no valid wrapping.

This example led Manuel Abellanas to pose the following more specific version of the conveyer-belt problem: do equal-size disks always have a proper band wrapping? This question has tantalized many researchers, and many attempts have been made to prove that the answer is 'yes', though so far all have failed. Still, most conjecture that the answer is 'yes', despite our lack of algorithm or proof.

MATHEMATICAL QUEST

The authors met at M.I.T. in 2007 to discuss the conveyer-belt problem. We followed a common technique in mathematics of exploring special cases or easier variations to make partial steps toward a larger solution to the whole problem. We invented two interesting variations to the problem and made partial progress on each.

Our first variation allows the addition of extra ('Steiner') disks. How many disks do we need to add to a given arrangement of disks to guarantee that all disks together have a valid band wrapping? It is relatively easy that, given n disks, we might need to add at least $\Theta(n)$ disks: just repeat an example like figure 26 $\Theta(n)$ times.[1] What would be interesting is if this is roughly the worst case:

Conjecture: Every arrangement of n disks can be augmented by $\Theta(n)$ additional disks so that the resulting arrangement has a valid band wrapping.

We make this conjecture because the following algorithmic approach seems promising, though we have not yet been able to formalize it into a solution. Start from a travelling salesman tour, a closed path that visits each disk exactly once, but may not be taut. One way to compute such a tour is to compute the visibility graph of the given disks (which disks can see each other, unobscured by other disks), compute a spanning tree of that graph (a minimal set of visibility connections that form a connected network), and take an Euler tour of this tree (walking around the tree) but avoid revisiting a disk by taking detours around it. Now the idea is to turn this tour into a valid band wrapping by adding tiny disks at key locations to effectively navigate the band where desired. The details of this process remain vague, but they seem feasible.

The bigger challenge is to determine how few disks can be added. We can afford to add a few disks for every turn in the tour, but the worry would be that navigating tight gaps between disks could require more than just a few disks. Nonetheless, we believe that this challenge is surmountable, at least for some tour. The conjecture should also be easier to prove for equal-size disks. At worst, it should be relatively easy to find an algorithm adding $\Theta(n^2)$ disks, because navigating a gap among n disks should require at most $\Theta(n)$ added disks, and the tour requires only $\Theta(n)$ such navigations.

Our second variation is to suppose that all disks are both the same size and 'separated', for an appropriate definition of separation. The idea for such a definition is to require that every two disks can see each other (in some sense) without unobstruction by other disks. In the weakest form, we can require just that there is a visibility line between the two disks; or we can require that every point on half of one disk can see some point on the other disk; or, in the strongest form, we can require that every point on half of one disk can see every point on half of the other disk. In these situations, or similar restrictions, it may be possible to follow a tour without adding any extra disks. Indeed, Manuel Abellanas has shown this result for the strongest of the three definitions of separation, and for any tour. This gives a positive solution to a special case of the equal-size conveyer-belt problem.

In the long term, our plan for this second variation is to pursue a hierarchical solution to the equal-size conveyer-belt problem, whereby we treat tightly packed clusters of disks separately from separated clusters of disks. We feel that these two extremes pose different challenges, each surmountable by themselves, hopefully in a way that can be combined. With separated clusters of disks, we can hope to use a solution to our second variation. With tightly packed clusters of disks, there are rather severe restrictions on how equal-size disks can be placed, and our hope is to exploit these constraints to find wrapping algorithms. This third variation is rather vague and probably the most difficult step in our plan.

PUZZLE AND FONT DESIGN

Having made some (but not a lot of) mathematical progress, we turned to challenging each other with puzzles along the lines of the conveyer-belt problem. The puzzle designer stuck identical push pins (representing equal-size disks) into a cork board, and the puzzle solver had to wrap a rubber band validly around the pins. To make the puzzle more challenging, however, the solution had to have one additional property: that it formed an English letter or word. The puzzle solver did not know which letter to aim for; the designer of course had one in mind, and aimed to

ensure that the answer was unique. This game quickly led to a series of puzzles and designs for making every letter and digit of the English alphabet. Figure 28 shows our preferred designs.

28 Our conveyer-belt alphabet, and the underlying disks.

The solved designs with the bands can be used as a new font with a mathematical backstory. Alternatively, the puzzle designs without the bands can be used as a 'secret code' that is most easily readable by those familiar with the mathematical problem.

29 Figure 29 provides a simple coded puzzle for the reader.

---

1 Here $\Theta(n)$ is asymptotic notation for 'a function growing linearly in n', or more precisely, some function that is between $a \cdot n$ and $b \cdot n$ for some constants $a, b > 0$.

References

[Abe08] Manuel Abellanas. 'Conectando puntos: oligonizaciones y otros problemas relacionados.' *Gaceta de la Real Sociedad Matematica Española*, 11(3):543–558, 2008.

[DD03] Erik D. Demaine and Martin L. Demaine. 'Hinged dissection of the alphabet.' *Journal of Recreational Mathematics*, 31(3):204–207, 2003.

[DD08a] Erik D. Demaine and Martin L. Demaine. Computational Origami. Permanent collection, Museum of Modern Art, New York. Originally part of Design and the Elastic Mind exhibit, February–May 2008. http://erikdemaine.org/curved/

[DD08b] Erik D. Demaine and Martin L. Demaine. Puzzles. http://erikdemaine.org/puzzles/

[DDEFF05] Erik D. Demaine, Martin L. Demaine, David Eppstein, Greg N. Frederickson, and Erich Friedman. 'Hinged dissection of polyominoes and polyforms.' *Computational Geometry: Theory and Applications* 31(3):237–262, June 2005.

[DDHPT08] Erik D. Demaine, Martin L. Demaine, Vi Hart, Gregory N. Price, and Tomohiro Tachi. *How paper folds between creases*. Manuscript, December 2008.

[DDLS05] Erik D. Demaine, Martin L. Demaine, Jeffrey F. Lindy, and Diane L. Souvaine. 'Hinged dissection of polypolyhedra.' In *Proceedings of the 9th Workshop on Algorithms and Data Structures*, Lecture Notes in Computer Science 3608, August 2005, pages 205–217.

[DDL99] Erik D. Demaine, Martin L. Demaine, and Anna Lubiw. 'Polyhedral sculptures with hyperbolic paraboloids.' In *Proceedings of the 2nd Annual Conference of BRIDGES: Mathematical Connections in Art, Music, and Science*, July 1999, pages 91–100.

[DDP06] Erik D. Demaine, Martin L. Demaine, and A. Laurie Palmer. 'The Helium Stockpile: A collaboration in mathematical folding sculpture.' *Leonardo* 39(3):233–235, June 2006.

[KDD08] Duks Koschitz, Erik D. Demaine, and Martin L. Demaine. 'Curved crease origami.' In *Abstracts from Advances in Architectural Geometry*, September 2008, pages 29–32.

30 **SCARLET ON TRAMPOLINE**

ANDREW CHAPMAN, CROSSWORD DESIGNER

AN ELASTIC PUZZLE

INSTRUCTIONS

Language is elastic—words stretch both their forms and meanings, and sometimes snap back to surprise us.

In this puzzle, both the structure and the content show elastic properties. The small grid is accompanied by a set of simple clues (mostly synonyms—question marks indicate double meanings). Each leads to an answer, the last letter of which begins the next word in line, rather like a crossword but without the need for numbers. The pink squares show the letters which are shared by adjacent words. Start writing in the answers at the top left, and follow clockwise in a spiral.

The second grid results from stretching the form and the content of the first. This time, the clues are cryptic. The answers are of course different words—but each has a formal relationship with the corresponding answer in the first grid, using its letters at the beginning and the end. For example:

If ART were in the first grid (though it isn't), ARRANGEMENT might be the answer in the second. Or PITY might stretch to PLASTICITY.

So, every answer in the second grid adds letters to the middle of the appropriate answer in the first (and solving an answer in one grid will help solve its partner in the other). Once again, the pink squares show where each word ends and the next one begins, and it all starts at the top left. One final note: all of the answers in the second grid are words that appear in other articles throughout this book, generally related to the main theme. A little something to stretch the mind.

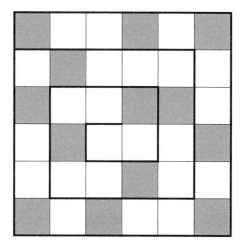

Puzzle answers

(capitals show the answers for the first grid)

1. LI(game)NT
2. T(hre)AD
3. DI(sto)RT
4. T(itani)UM
5. ME(chanic)AL
6. L(iqu)ID
7. DE(velopme)NT
8. T(echnoscientifi)IC
9. CU(ltu)RES
10. S(tretc)HED
11. D(eformati)ONS
12. S(truct)URE
13. EL(asticit)Y

CLUES
(1st grid, then 2nd grid)
1 FLUFF (A melting, disintegrating tissue) 2 SMALL AMOUNT (Thaddeus swallows hollow rye for fibre) 3 SOIL (Falsify Diana's legal wrongdoing) 4 STOMACH (Metal-scarred mutant with two extra eyes) 5 REPAST (One of Shakespeare's amateur actors might be wooden?) 6 COVER (51 pounds of fluid) 7 IMPRESSION (Ruined Devon temple becomes building site) 8 TWITCH (The

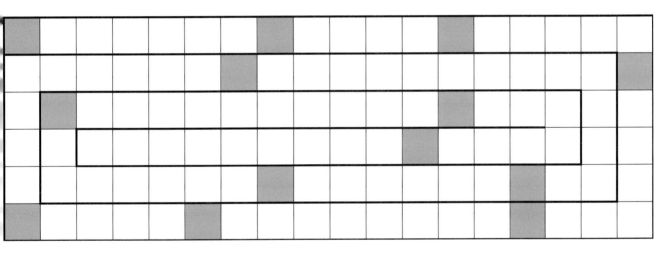

Tissue Culture & Art Project's work is suffering hectic infections?) 9 SUCCESSFUL TREATMENTS (Extremist movements engulf ancient city next to Spain with particular ways of life) 10 LOST SHELTER? (Pulled and carried off without hesitation) 11 ACADEMICS —OR MAFIA? (Assaults on beauty: science fiction in moderation, perhaps) 12 INDISPUTABLE (Composition is certain to contain French gimmick by Tissot) 13 CITY IN ENGLAND (& Nevada) (Spring is late, maybe to Paris, for example)

A LINE FOR WATER

All morning we spoke
the imagined was probable

future but also mentioned
specifics like freedom of hair

design and the art of kissing the cheek

We drove around the neighborhood toilet
paper ribboned the live oak

we stopped by the distressed
magnolia as a town

home was being
built by its north face

the earth was thin over the deep root

I left the car running fuming
stepped out to see the tree was blooming

on all heights all around
let's go back you said

before more ice melts and the polar bear
would fight the grizzly

and what a documentary
that would make

But radio stations don't keep
the same frequency across state lines

the colonel said things other colonels say
they marched were marched

through maize fields sucked

on maize but the maize was dry
and thirst was donkey piss thirst

A boy didn't know how to swim
jumped in a well his mother told him

this would be the day she'd die

and parched they pulled him out
sucked his clothes then marched were marched

a boy asked the guard for some water

a suffering succotash martial art
duel frame by frame

into seizure and never came back

We'd had enough of the interior
the bathroom photo of children

frolicking in a fountain
in a city where mothers vote

got our family picture taken our hair cut
our subdivision's catch-and-release

pond given a fountain two
pairs of ducks and herons

we tried to capture in flight on camera
the dumb fish swimming about

with punctured lips
having not learned from their previous errs

A poet was a soldier once
and is now forgiven

a chicken crossed the road to listen
to some Bach Bach you said and laughed

and desert is one of water's names
showers from wells we dig

skin cracks that stomp the ground
'til God blows up

an aquifer beneath the feet
soda orange water the sand had settled in

or another untouchable soil
let loose from the conquered hills

animals killing farms
women and young girls jump

up and down the water pump
handle an oil drill

smiling how beautiful their toned arms

We rode through the fountain
on leaf blower mornings

walked in the heat collapsed
and needed to be hung

to drips and were
wild in the movies

we swapped our strollers for slings
and were jigsaw-puzzled on the thoroughfare

and the baby with the bath water
we bibbed were bibbed

ran loads in the washer
love's no policy the think-tank man said

over dinner and wine
and wine is a diuretic

A boy walked back to his village
alone carrying nothing in the night

wolves howled and guns cocked

a mother toiled the land illiterate a world spun
and was round as an ant

or a pebble was set on a plate
gravity holding place

our eardrums flapped as tarp
that feeling

a mirage a boy
on a bicycle through checkered

tree shadow clearing
the fog with a machete

A boy out of his upper-end
house with a wooden sword

headed for the bamboo
that landscapes his house

we were strolling by then
you were taken

by trees
an ancient magnolia

a giant lemon blossom you called it

STRETCHING THE POINT

After nearly two years of 'critical' debates, tonight was Barack Obama's most impressive performance yet. Not only did he show an unbelievably elastic knowledge of just about every issue covered in every newspaper over the past five years, but his cadence, delivery, thoughtfulness and overall make-up were stunning.

Barack Obama demonstrated 'unbelievably elastic knowledge' during his debate with McCain (Second Debate, October 7, 2008)
http://ccsbandwagon.blogspot.com/2008/10/second-debate.html.

A SNAKE CAN COIL AND STRETCH BUT IT CANNOT OUTRUN ITS HEAD

LOVE AFFAIRS AND BANKS

In most popular usage, elasticity is more related to the physical property of an object—its capacity to return to its original form after being s-t-r-e-t-c-h-e-d. A rubber band or bungee cord is elastic. It is important to note that there is a *limit* to how far a rubber band can be stretched. Beyond that, it will snap and contract to their original form or structure. Thus, elasticity has a breaking point.

The concept can also be applied to ideas, thoughts, practices or any other concept that has the capacity to adapt to new situations or the capability of being 'stretched'. For example, you can only put so much weight on a bridge; its rigid structure is inflexible. On the other hand, a love affair may be considered 'elastic' and can accommodate various demands and exigencies but up to a certain point. You can push a wife only so far before she snaps. Then comes a dirty look—or the frying pan, depending upon the culture! Conversely with a husband.

Over centuries of human experience, definitions, concepts and the meaning of certain things in life have assumed various degrees of 'elasticity' and been pushed to their limits—often in defiance of common sense. A love affair can result in an abusive relationship when one partner takes advantage of the other. The meaning of laws and regulations can be stretched beyond their limits and abused. A favourite is the 'non-bank banks' of Wall Street. These are the investment banks that lend large sums of money to corporations. As such, they are banks but because they do not accept deposits from the public, they are not strictly banks subject to banking regulations. Hence, they are non-bank banks. Confused?

The current stock market bubble, the mortgage crisis of 2008 that sparked a financial meltdown and a global recession are recent examples of abuse of regulations and taking things to the limits. Rogue hedge fund managers 'pushed the envelope', investing massively in speculative derivatives and credit default swaps, cranked out with complex mathematical formulae and computers. Companies were 'over-leveraged'. Banks s-t-r-e-t-c-h-e-d the notion of 'viability' with sub-prime mortgage rates to home-owners who did not have the ability to pay. Bernard Madoff took the concept of trust to new depths of criminality in a $50 billion Ponzi scheme.

Some time ago, company executives earned hefty bonuses when their companies performed well. Today, the term 'bonus' has become very 'elastic'. Company executives earn bonuses even when their companies are collapsing and are in need of a government bailout. In December 2008, Merrill Lynch paid $3.6 billion in bonuses to its executives after losing more than $27 billion for the full year.[1]

Common sense—can only be defied up to a certain point as elasticity has its breaking point. Bubbles burst and the 'frying pan' will come down with metaphysical certitude. The US Congress has capped the salaries of company executives who receive government bailout money at $500,000. There will be 'market correction' but true to the flaws of human nature, there will be new bubbles. To err is human and to repeat them—well, that is also human.

Beyond the stock market and executive bonuses, there are several other things whose definitions have become very 'elastic': 'government', 'democracy', 'justice', 'freedom of expression' and even 'tolerance'. Nobody knows what 'human rights' stand for any more. In 2007, countries that sat on the United Nations Human Rights Commission included Libya, North Korea and Zimbabwe. Zimbabwe even chaired the United Nations Commission on Sustainable Development in 2008! These days, freedom of expression is abused. It is misinterpreted by some as license to spit venom at other groups, who are supposed to absorb it and demonstrate that they are not 'bigots' or 'racists.' But even this freedom has its limits. Try shouting "Fire!" in a crowded theatre or "Hijack!" on an airplane.

RELIGIOUS INTOLERANCE AND IMPERIALISM

In the aftermath of September 11, the interpretation of religious tolerance has become very elastic in only one direction. It is obligatory to be tolerant of other people's culture and religious beliefs but not them of yours. Muslims want to be tolerated in Christian countries but are they of Christians in their own Muslim countries? This double standard or hypocrisy has created enormous problems worldwide, especially in Africa, where religious intolerance has been pushed beyond the absurd into lunacy. Neither Islam nor Christianity is indigenous to Africa. So why should Islam be forcibly be imposed on Africans and why should certain African countries be declared Islamic states?

Historically, Africa has been hospitable to foreigners and their religions. But the welcome mat was trampled upon in eager abuse of hospitality. The main culprit has been Islam, which was originally introduced to Africa by Arab traders, conquerors and slave raiders. Christianity was introduced by European traders, conquerors and slave raiders. Thus, strictly from the black African historical perspective, the Arabs were no different from the Europeans. Both groups were invaders, colonizers and slavers, who used their religions—Christianity and Islam—to convert, oppress, exploit and enslave blacks.

While the Europeans organized the West African slave trade, the Arabs managed the East African and trans-Saharan counterparts. Over 20 million black slaves were shipped from East Africa to Arabia. Enslaving and slave trading in East Africa were peculiarly savage in a traffic notable for its barbarity. Villages were razed, the unfit villagers massacred. The enslaved were yoked together, several hundreds in a caravan, on their long journey to the coast. It is estimated that only one in five of those captured in the interior reached Zanzibar. Some historians believe the slave trade was more catastrophic in East Africa than in West Africa. For the trans-Saharan slave trade, an estimated 9 million captives were shipped to slave markets in Fez, Marrakesh (Morocco); Constantine (Algeria); Tunis (Tunisia), Fezzan, Tripoli (Libya); and Cairo (Egypt). Yet few Arabs acknowledge this horrific part of African history.

After independence in the 1960s, a new wave of religious imperialism emerged in Africa. In Mauritania and Sudan, blacks found themselves under Arab masters and the enslavement of blacks by Arabs continues to this day. On the genocide against blacks in Darfur, the United Nations accused Sudan's government of encouraging the attacks by Arab militias against blacks in what officials call a "campaign of ethnic cleansing".[2] Said Aloysius Juryit of Nigeria: "Events in the Sudan and Mauritania (to mention only a few) have shown that the worst racists are Arabs, especially when it comes to dealing with blacks".[3]

Elsewhere in Africa, ten northern Nigerian states have imposed the *sharia* in clear violation of Nigeria's own constitution. Terrorist bombings have occurred in Kenya, Tanzania and South Africa. Crass attempts are being made to impose Arabic names and Islamic law. According to the Amazigh (Berber) Cultural Association in the U.S., a new Moroccan law, enacted in November 1996 and referred to as Dahir No. 1.96.97, "imposes Arabic names on an entire citizenry more than half of which is not Arabic." But tolerance and hospitality have their limits.

In Algeria, the Berbers are fighting back. Fed up with years of discrimination and persecution at the hands of the Arab masters, Berbers, who make up 20 percent of Algeria's 32 million people, seek more autonomy in the eastern region of Kabylie. They were the original inhabitants of North Africa when invading Arabs introduced Islam in the seventh century. Old tensions erupted into violence after a Berber schoolboy died in police custody in April 2001. There were street clashes in Kabylie between the police and Berber militants, and more than 100 protesters were killed.

Elsewhere, black Africans have been venting their spleen at the string of senseless terrorist bombings:

Kenya and Tanzania in August 1998 that claimed more than 240 lives,

On August 26, 1998, Islamic terrorists blew up the Planet World restaurant in Cape Town, South Africa, killing one person and injuring 27.

In November 2002, Islamic terrorists struck again with an attack on Paradise Hotel in Mombasa, killing at least 30 people.

If the Islamic terrorists thought they could count on black Africans for sympathy or solidarity while using them as cannon fodder for their cause, they terribly miscalculated. They only succeeded in shattering the crucible of Afro-Arab solidarity and purchasing an excess supply of black African wrath in the bargain. The twin bombings in East Africa blew the lid off anti-Arab rage. Said an irate Nigerian medical doctor, Segun Tonyin Dawodu: "Why on an African soil? Damn the stupid imbeciles. The OAU and other African Organizations should condemn these unprovoked atrocities against black people. All Arabs should immediately be rebuked without mincing words and there should be a blanket ban on issuance of visa for entry into any African country by these bigots."[4]

The angry African reaction seeks to invoke the principle of reciprocity. Those who preach or seek religious tolerance should themselves practice or extend it. And those who claim to have a 'superior' religion should demonstrate it. After all, a religion should be about saving people regardless of their race, ethnicity, sex or creed—not blowing up or beheading non-believers.

GOVERNMENT

Governments the world over have also sought to 'push the envelope', expanding enormously their role in society and encroaching on individual liberties—all ostensibly to 'protect the people from themselves'. Nearly every conceivable aspect of life is regulated by governments that live beyond their means.

In the process of regulating and taking away people's freedoms, governments rack up massive deficits, borrow recklessly to finance profligate spending, and accumulate huge debts for future generations. The US national debt stands $10.85 trillion as of February 24, 2009. That is the sum of all federal deficits (occasionally reduced by rare budget surpluses) since Alexander Hamilton restructured the Revolutionary War debt in the late 18th century.

The US national debt is financed through the sale of US, Treasury Bills and Bonds, 44 percent of which are held by foreign investors. Guess who holds the largest stock of US Treasury Bills? Communist China, which owns about $700 billion of US government debt! The dirty, little-known irony is that it is a communist country that is rescuing the champion of capitalism.

On her Feb 2009 trip to China, US Secretary of State Hillary Clinton encouraged China to lend more to the US Government: 'I certainly do think that the Chinese government and central bank are making a smart decision by continuing to invest in [US] Treasury bonds.'[5] Smart decision? How about the US government making the *really* smart decision of living within its means?

One government that has debased its currency by over-printing money to finance irresponsible fiscal excesses to the point of insanity is that of Zimbabwe. Its currency is now worthless; its 10 billion Zimbabwean dollar note is worth only 3 cents. In 2008, its rate of inflation reached a staggering 231 million percent—whatever that means. It even ran out of paper on which to print the currency! Then the government tried to tame hyperinflation—by banning price increases!

Elsewhere in Africa, the notion of 'government' has been stretched to the point of absurdity. Centuries ago, government was created to protect and serve the people. But in post-colonial Africa, this function has been turned completely on its head. Government doesn't serve the people; it preys upon and fleeces them. A proliferation of vampire states, coconut republics and failed states dot the African continent. The rule of law is a farce. Bandits are in charge and their victims in jail. The police are highway robbers and judges are themselves crooks who protect the ruling bandits. A cabal of gangsters, who use the government machinery to enrich themselves, their cronies and tribesmen, has hijacked the state. The chief bandit is the head of state himself. Said Simon Agbo, a farmer in Ogbadibo, south of Makurdi, Benue state capital in Nigeria, "I heard that we have a new government. It makes no difference to me. Here we have no light (electricity), we have no water. There is no road. We have no school. The government does nothing for us."[6] In Angola, a pamphlet of *Parti d'appui démocratique et du progrès d'Angola* (PADPA) scolded:

> Thousands of Angolans are dying of hunger because the country is mismanaged and the holders of power have turned into a band of thugs who pretend to be managing a bank. Our bank. Our petrol. Our diamonds. Our riches. But above all, our children, parents, brothers and cousins, who they use as fodder for their diabolical cannons.[7]

But here too, Africans are fighting back against predatory vampire states. In Senegal, The Loulouni district administrator was thrown out of the village when he tried to collect taxes on Feb 2, 1995. He returned on February 9 with a battalion of police and paramilitary gendarmes. Enraged villagers met them with clubs and hunting rifles. Two peasants and eight policemen were wounded in the ensuing clash.

Back in 1991, Amina Ramadou, a peasant housewife, had a solution to her country's (Zaire's) economic crisis: "We send three sacks of angry bees to the governor and the president. And some ants which bite. Maybe they eat the government and solve our problems."[8] Unfortunately, they sent 'angry' rebels who 'ate' the country in 1996. The new country is now called the 'Democratic Republic of Congo'.

DEMOCRACY

This is another very elastic concept, stretched by crackpot regimes to cover up their tyrannical excesses. The 'People's Republic' of this, the 'Democratic Republic' of that—as if the people have any say in the running of the affairs of the country. Such countries hold farcical 'elections' in which the incumbent writes the rules of the game, and then serves as a player, the referee, and the goalkeeper all at the same time. Imagine. The deck is hideously stacked against the opposition candidates, who are starved of funds, denied access to the state-controlled media, and are brutalized by government-hired thugs, as the police watch. Opposition parties are banned and their leaders tossed into jail or in hiding.

By contrast, the incumbent enjoys access to enormous resources: state media, vehicles, the police, the military and civil servants—are all commandeered to ensure his re-election. Further, the entire electoral process itself is rigged: voter rolls are padded with ruling party supporters and phantom voters, while opposition supporters are purged. The Electoral Commissioner is in the pocket of the ruling party, as are the judges who might settle any election disputes. During the election campaign, posters of the incumbent are everywhere, while pro-government thugs terrorize the populace and anyone perceived to be a supporter of the opposition parties. Innocent civilians are force-marched to attend incumbent party's rallies, while opposition rallies are violently disrupted and opposition supporters brutalized and even killed, as the police look on.

Naturally, the incumbent 'wins' 99.99% of the vote to declare himself 'president for life. 'Witness recent 'elections' in Ethiopia (2005), Nigeria (2007), Kenya (2007) and Zimbabwe (2008).

Here is a jolting caricature of 'President Elasticity' I created in 2004:

> **I am Musugu Babazonga, the President-For-Life of the Coconut Republic of Tongo in the Gulf of Guinea. Don't mind Julius Nyerere of Tanzania; he called himself 'Mwalimo' (Teacher), while Mobutu Sese Seko of Zaire changed his name to 'Sese Seko Kuku Ngbendu Wa Za Banga', which, in the local Lingala language, meant, 'The rooster who leaves no chicken untouched'. And forget about Idi Amin who called himself 'The Conqueror of the British Empire'. My name trumps them all. My people call me the 'Cutlass' because I behead terrorists. Anyone who opposes my rule is a terrorist. They give me nightmares. Haba. So, I am also fighting a war against terrorism.**
>
> **I agree with everyone that rule of law, free and fair elections, transparency, accountability, stability, and foreign investment are all important in the process of economic development and rejuvenation. But we have all these in my country. It is an insidious form of racism and imperialism to claim that we don't.**
>
> **I wrote the laws of the country myself and, since I am the ruler, we have the rule of law. I have been in power for 30 years, so we have stability. If US Secretary of State Colin Powell calls my regime non-democratic and totalitarian, I will sue him for defaming and humiliating my people [Libya].**
>
> **We have just concluded our first elections in 30 years and they were 'free and fair'. Those who opposed me are in jail, where they are free to say what they want [Togo]. Nobody bothers them there. I think that's fair. I won 99.92% of the vote. In fact, I would have won 110% if the opposition had bothered to get out of their graves to vote for me. Lazy bunch. So, you see, our elections were free and fair.**
>
> **Yes, I take 'development' very seriously. My pockets are well developed. We have 'foreign investment' too. My wealth is safely invested in Riggs National Bank in Washington, D.C. and other foreign banks to protect against foreign exchange fluctuations [Obiang of Equatorial Guinea]. Yes, my country produces oil but the oil revenue is a state secret—to protect it from the prying eyes of imperialist enemies. My finance minister keeps the accounts at a secure place—in a coconut tree [Zambia].**
>
> **World Bank officials said we need less government spending so we are feverishly working on that. We have set up a Ministry of Less Government Spending [Mali]. The Bank also said we should privatize state-owned enterprises, so I sold them to my relatives and friends, who are in the private sector [Egypt, Kenya, Nigeria, Uganda and Zimbabwe].**
>
> **And I don't play with 'accountability'. My political opponents must account for every penny they spend and explain where they are at any moment of time. My security forces have been instructed to verify that**

information and must report to me every step my critics and political rivals take [Eritrea, Ethiopia, Togo, Zimbabwe]. I know who they are and where they live. So we have 'accountability'.

We have 'checks and balances' too. All government officials are required, by decree, to have bank accounts on which they can write checks and their bank accounts must have positive balances. Bounced checks are not tolerated. So we have checks and balances.

President Bush says his Millennium Challenge Account (MCA) will give aid only to those governments that govern justly, promote economic freedom, and invest in the people. No problem there. In fact, my country should be the first to receive aid under that program. I am the Constitution; I wrote it myself and set myself a two-term limit before our first elections. All the years I have held power before the Constitution came into effect didn't count [dos Santos of Angola and Rawlings of Ghana]. It is 'just' because I have done a lot for my country. And if I don't like the two-term limit, I will change it [Deby of Chad, Conte of Guinea, Nujoma of Namibia, and Museveni of Uganda]. And when I retire, my son will take over [Mubarak of Egypt]. He will be freely chosen by me. So, you see, we have constitutional rule and I govern 'justly'. All the senior positions in my government are filled with my relatives, friends, and tribesmen [Burundi, Cameroon, Kenya and Rwanda]. They love me, go ask them. We have two radio stations in the country and I own them. People can say what they want only on my radio stations because freedom of expression is an outmoded concept [Jonathan Moyo of Zimbabwe].

Movie theaters and television are forbidden to show pornographic material, which corrupts the minds of the young [Bashir of Sudan]. Violators are beheaded; they can appeal later. So, we are aggressively combating corruption of children. And we do actively promote economic freedom. My ministers can engage in whatever economic activity they like.

There is no famine in my country. The people are well-fed; they eat grass. Hey, cows eat grass too. Unfortunately, there are no cows left in the country because the people ate all the grass. If we are poor, it is because of Western colonial plunder and exploitation. So the West owes us—big time; in fact, $777 trillion! Actually, that figure was adopted by the African World Reparations and Repatriation Truth Commission in August 1999 in Accra, Ghana (the 1999 Accra Declaration) and signed by Dr Hamet Maulana and Mrs Debra Kofie, co-chairpersons of the commission. We are waiting for the cash. (From *Africa Unchained*.)

This is utter debauchery of the concept of elasticity—s-t-r-e-c-t-c-h-e-d to the point of lunacy. Would somebody please smash this coconut-head? Africans are determined to do so.

An increasing number of them—including even children—are voicing their outrage at the scandalous failure of African leaders to bring development to the continent. At the United Nations Children's Summit held in May 2002 in New York, youngsters from Africa ripped into their leaders for failing to improve their education and health. "You get loans that will be paid in 20 to 30 years… and we have nothing to pay them with, because when you get the money, you embezzle it, you eat it", said 12-year-old Joseph Tamale from Uganda.[9]

An irate Horace Awi, a member of the Concerned Professionals Group and a drilling engineering manager with a multinational oil company in Lagos, Nigeria,

wrote on a naijanet discussion forum on November 16, 2001:

> The more you read about Africa, the more it becomes evident that African leaders are a strange lot. These guys are worse than space aliens. And somebody wants me to believe our problem is the white man. Rubbish. I posit that colonial rule was better. Obasanjo, the Nigerian leader regards himself as the best black leader in the world today. Maybe Mandela is white. This is why Obasanjo gallivants all over the globe. Let's concede that perhaps he is. Then Africa is really in trouble. If the best rules like they are doing in Nigeria today, frittering away our poor income on nonsensical projects, you begin to wonder what hope the African has?
> (Quoted with permission).

In August 2001, the Sierra Leonean government tried urging people to stop jeering and throwing stones at former military ruler Captain Valentine Strasser, who became Africa's youngest head of state when he seized power at the age of 25 in 1992 and was overthrown in bloodless coup in 1996. "A government statement said Captain Strasser had been embarrassed by people throwing stones at him and booing him when he ventured out on the streets of the capital, Freetown."[10]

Former South African president Nelson Mandela weighed in, urging Africans to take up arms and overthrow corrupt leaders who have accumulated vast personal fortunes while children have gone hungry. He urged the public to pick up rifles to defeat the tyrants. And no less a person than Nobel laureate Archbishop Desmond Tutu added his voice. In an interview with the *Saturday Star* newspaper in Johannesburg, he said: "Robert Mugabe of Zimbabwe seems to have gone bonkers in a big way. It is very dangerous when you subvert the rule of law in your own country, when you don't even respect the judgments of your judges then you are on the slippery slope of perdition. It is a great sadness what has happened to President Mugabe. He was one of Africa's best leaders, a bright spark, a debonair and well-read person."

THE BREAKING POINT

Elasticity has a breaking point which is eventually reached. A day of reckoning eventually arrives. Market bubbles burst. Governments go broke. There is a limit to how much pain, suffering and oppression the people can endure. As well, tolerance has a breaking point.

A year after taking office, Niger's president Maharanee Ousmane had tripled his personal fortune. As required by law, President Ousmane had declared a fortune of 51 million CFA ($89,000) and three houses when he took office in April 1993. A year later, "The poor West African country's Supreme Court said on April 28, 1994, that Maharanee had declared 160 million CFA ($280,000), with 57 million CFA held in cash and the rest in a local bank. Maharanee's list of property was 10 houses in Niger, livestock and poultry, three cars, two television sets, two video recorders and two gold watches."[11]

He built a huge security apparatus to protect him. His Presidential Guards were hulky and fearsome.

They could crack a living skull with bare hands. But on Jan 27, 1996, Ousmane was overthrown in a coup led by General Ibrahim Bare Mainassara. On the day of the coup, Ousmane's Presidential Guards fled. A gaggle of them dove into the Niger River to escape. But the crocodiles got them. Oh what a hearty feast they had.

The next buffoon to assume power didn't learn. Being a product of that security structure, with intricate knowledge of its inner workings, General Mainassara repaired the weaknesses and strengthened the structure—at least, that was what he thought. Under pressure from Western donors, General Mainassara called for elections on July 6, 1996 and decided to contest the presidential elections himself. When early results showed that he was losing,

> Mainassara sacked and replaced the Independent National Electoral Commission (CENI) with his own appointees, placed his opponents under guard in their own houses. The other contenders' home phone lines were also cut off. A ban on public gatherings in Niamey was announced on the evening of 9 July. Security forces were deployed at candidates' homes and some political party offices. The floodlit Palais des Sports where results were centralized was guarded by an armoured car and heavy machine guns mounted on pickup trucks. Two radio stations were stopped from broadcasting and all of the country's international phone lines were suspended.[12]

Having committed such grotesque travesty, he was ill at ease. He didn't trust his own military, so he created a Special Presidential Guard and fortified his palace. It was impregnable but just in case somebody might have an idea of attacking from the air, he gave his Presidential Guard some heavy-duty artillery, including powerful anti-helicopter machine guns.

Sometime in 1999, returning from a trip overseas, his Presidential Guard went to the airport to meet him. They decided to 'test' their weapons and opened fire with their anti-helicopter machine guns. Mainassara's body was shredded into pieces, littering the tarmac. His wife, upon seeing the bits, collapsed on to the tarmac.

Security and other concepts may be elastic but they cannot be s-t-r-e-t-c-h-e-d beyond the point where they defy common sense. Says an African proverb: 'A snake can coil and stretch but it cannot outrun its head.' Somalia, Rwanda, Burundi, Zaire, Liberia, Sierra Leone, Ivory Coast, Togo and Sudan all imploded and descended into civil African Republic, Chad, Egypt, Equatorial Guinea, Eritrea, Ethiopia, Guinea, Libya, Uganda, and Zimbabwe.

1   http://cbs13.com/national/Merrill.Lynch.bonuses. 2.941762.html
2   *The New York Times*, July 8, 2004; p.A4
3   *New African*, March 1990, p.6
4   naijanet@esosoft.com, August 8, 1998
5   *The Washington Times*, Feb 23, 2009; p.A15).
6   *The Washington Times,* Oct 21, 1999; p.A19
7   *The Economist*, Feb 3, 2001; p.47
8   *The Wall Street Journal,* Sept 26, 1991; p. A14
9   *BBC News* website, May 10, 2002
10 *The Daily Graphic,* Aug 18, 2001; p.5
11 *African News Weekly*, 20 May 1994, 8
12 *African News Weekly*, 15–21 July 1996, 2

References:
Ayittey, George B. N. (2005). *Africa Unchained: The Blueprint for Africa's Future.* New York: Palgrave/MacMillan.
Carabini, Louis E. (2007). *Inclined to Liberty.* Auburn, AL: Ludwig von Mises Institute.
Lamb, David (1985). *The Africans.* New York: Vintage Books

THE EVER CHANGING
MARGINS
OF THE LAW

Human rights are considered to be innate in every human being; they are the rights one has simply because one is a human being. They are those rights needed to be able to live a life of human dignity. Hence it is not surprising that these rights are said to be 'fundamental'. One might expect these rights to be clearly defined and neatly circumscribed. However, that is only true to a limited extent and several (often inter-related) instances of 'elasticity' of these rights can be identified.

First of all, the categories of rights that are qualified as human rights expand over time, revealing that the boundaries of the concept can be developed and reinterpreted. Secondly, the interpretation of particular human rights is not static but evolving, entailing (mostly) an expansion of the protected right. This expansion can occur because of the widening of the scope of application of the right and/or the diversification of the state obligations and/or the reduction of the limitations to the effective enjoyment of these rights that are considered acceptable (ie legitimate).

In the wake of 9/11 and perceived threats of terrorist attacks, several states (and the EU) feel the need to and feel justified to encroach further on fundamental rights—particularly the right to respect for privacy and the right to personal security (e.g. protection against undue incarceration). Whereas International courts tend not to tow this line of thinking, the response of national courts is split, with some going along

i

with this development more than others.

Human rights are enshrined not only in national constitutions and legislation but also in several international conventions. In addition to the numerous conventions that have been developed in the UN framework, they have also been taken up in conventions of the Council of Europe, the Organisation of American States and the African Union. The following account will often draw on the jurisprudence of the European Court of Human Rights (ECHR) in relation to the European Convention on Human Rights of the Council of Europe because it is the most developed, richest and thus generally most highly respected jurisprudence. However, findings of other courts are also taken into consideration, especially when they reveal differences in interpretations and converging or diverging lines of argument.

Human rights are not confined to those relatively well-known ones such as freedom of expression, freedom of religion, the right to respect for one's privacy and the prohibition of torture. In addition to these civil and political rights, social and economic rights such as the right to housing and health care are also considered to be human rights. Since the late 1980s a distinct category of rights of peoples has been emerging, sometimes referred to as solidarity rights. Examples of the latter would be the rights of peoples to peace and to development.[1]

Some academics warn against the tendency to qualify ever more rights as human rights. The proliferation of human rights could lead to a detrimental corrosion of their status.

It is furthermore relevant to point out that while initially the differences between civil and political rights on the one hand and social and economic rights on the other were highlighted, from the mid 1990s this emphasis has shifted to an interdependence and indivisibility of human rights. Interdependence suggests that the full realisation of one of these rights needs or presupposes the effective enjoyment of the others. This also applies to the boundaries between civil-political and socio-economic categories. In addition, closer scrutiny reveals that in the end the state obligations flowing from these rights

ii

are not that different in nature after all.

For example, the right to respect for one's private life, family life and home has been understood by the ECHR to encompass the right to a healthy environment. Arguably when one is continuously exposed to fumes and toxins in the air or polluted ground water one is not in a healthy and safe home environment. The Inter-American Court of Human Rights has included several considerations of socio-economic rights in its jurisprudence not via the right to respect for privacy but via the right to respect for life: ie a human right. According to this Court the right to life should be understood as a right to a dignified life which in turn would presuppose adequate housing, food, sanitation etc.

Another manifestation of the interdependence of civil-political and socio-economic rights concerns the identification of positive state obligations in relation to civil-political rights. This had traditionally been in the domain of socio-economic rights. Civil-political rights were considered to entail negative state obligations of non-interference: states were not allowed to interfere disproportionately with the freedom of expression or the privacy or the freedom of religion of persons under their jurisdiction. The inclusion of positive state obligations in civil-political rights has repercussions for their scope of application. The positive obligations identified by the ECHR have allowed the Court to tread de facto on the terrain of social and economic rights, for example in the case of rights for disabled people to adapted housing, access to public areas, and access to legal aid in particular circumstances. While the Court has been careful and wary for these obligations not to have too far-reaching financial implications for the state parties to the Convention, the interpretative leap into the field of social and economic rights is no less real.

Positive obligations concerning civil-political rights include those of the state to ensure the respect of human rights in relation between private individuals. For example, the state has the positive duty to make sure that the right to respect for

iii

privacy is respected by one's neighbours and to some extent even one's employer. This implies the need to have suitable provisions of criminal and tort law. Other types of positive obligations have added procedural dimensions to fundamental rights. For instance, in regard to the right to respect for life, states have been obliged to ensure independent and effective investigations whenever there are instances of suspicious deaths.

Human rights are not static but develop constantly in line with changes in human society. This evolutionary nature of human rights is not only visible in the steady expansion of rights that are considered to be human rights as outlined above, but also in the development of how human rights are recognized.

Concerning shifts in interpretation, excellent examples can be found in the jurisprudence of the ECHR, which has developed its famous 'living instrument' doctrine in this respect. The European Convention on Human Rights is considered to be a living instrument, which needs to be interpreted in line with the changes and developments in society. According to the Court, the recognition of the evolutionary nature of human rights is essential for these rights to be real and effective and concerns both the interpretation of the field of application of the human rights and the scope that is allowed for legitimate limitations.

When one reads article 8 ECHR it stipulates the right to respect for one's private life, one's family life, one's home and one's correspondence. The case law of the ECHR has revealed that private life concerns a broad variety of issues ranging from protection against telephone tapping and searches of one's house to the official recognition of a gender change. The living instrument doctrine was used by the Court to justify the inclusion in the field of application of article 8 ECHR of the right to a traditional way of life for minorities. Furthermore, the right to respect for one's privacy extends, to some extent at least, to the professional sphere. There have not been many cases on this area but enough to conclude that this does

iv

not only apply to professionals that often work from home like lawyers (whose office can only be searched in limited circumstances)[2] but also to employees in a big firm (whose email cannot be systematically checked by their employer),[3] and even to the registered offices of a firm with legal personality.[4]

Shifts in interpretation have also been identified in relation to the scope of legitimate limitations to the effective enjoyment of human rights. Indeed, most human rights, however 'fundamental' they may be, are not absolute. With a few exceptions, such as the prohibition of torture, inhuman or degrading treatment or punishment, governments are allowed to limit the exercise of human rights. The doctrine of legitimate limitations concerns the criteria that need to be complied with for these limitations to be lawful. To the extent that states limit the exercise of fundamental rights without respecting these criteria (by transgressing the boundaries of the legitimate limitations) they violate the human rights concerned. An example in relation to the right to respect for privacy would be the search of the house of a suspected criminal. This would constitute an interference with his right to respect for privacy. However, to the extent that legislation regulating searches concerning probable cause, procedural requirements and proportionality has been complied with, this interference would not amount to a violation of the right concerned.

Remarkable shifts in jurisprudence have taken place in relation to the right to respect for one's private life. While the European Court used to accept that states criminalised sexual activities between homosexuals, since the 1980s it no longer accepts this when it concerns consenting adults. Change of one's gender is now also legally seen as an aspect of one's private life. The Court used to accept the argument by states that it would be too burdensome to demand that they make the necessary adaptations in their public registers so that post-operative transsexuals at all times would be treated in accordance with their new gender. Since 1986 (Rees v UK, 25 September 1986) the Court has indicated that it may hold

that the lack of adaptation of the official registers does not amount to a violation of article 8 just yet, but that "the Court is conscious of the seriousness of the problem affecting these persons and the distress they suffer. The Convention has always to be interpreted and applied in light of current circumstances… The need for appropriate legal measures should therefore be kept under review having regard particularly to scientific and societal developments" (paragraph 47). Only with the Christine Goodwin judgement of 11 July 2002 did the Court change its position. According to the Court at that time there was "clear and uncontested evidence of a continuing international trend in favour not only of increased social acceptance of transsexuals but of legal recognition of the new sexual identity of post-operative transsexuals… the Court considers that society may reasonably be expected to tolerate a certain inconvenience to enable individuals to live in dignity and worth in accordance with the sexual identity chosen by them at great personal cost" (paragraphs 86–90).

The latter developments in relation to the acceptable scope of the legitimate limitations is actually intrinsically related to another famous doctrine of the European Court of Human Rights, more particularly the one concerning the 'margin of appreciation' of states. According to this doctrine national public authorities are better placed than the European Court to determine the requirements that flow from human rights in the particular circumstances of a case, since these authorities have a more detailed knowledge and understanding of the relevant context, including 'local' customs and morals.

It is important to realise that the margin of appreciation goes hand in hand with European control; that is control by the ECHR. Still, it is obvious that the broader the margin of appreciation left to states, the narrower the level of scrutiny (in relation to the limitation) adopted by the European Court. Indeed the margin of appreciation and the level of scrutiny are two sides of the same coin.

Furthermore the margin of appreciation is not always

React

equally extensive. While the European Court has not yet developed a proper theory that is consistently applied, a few relevant factors to determine the exact width of the margin in a particular case have been identified by academics. One prominent factor concerns the extent to which a common European standard can be discerned in relation to a particular human rights issue. The existence and acknowledgement of a European consensus is itself subject to change, and hence the extent of the margin of appreciation similarly changes.

The degree of the margin of appreciation that is allowed to states is a clear manifestation of the elasticity of human rights. The wider the margin, the more states can decide for themselves what is acceptable and what is not in relation to interferences with the enjoyment of human rights. Hence, the greater the potential for differential levels of compliance with these rights between states.

Finally, in the current era of perceived threats of terrorist activities, it seems important to stress that the ECHR is careful not to allow states to impose too far-reaching limitations to human rights in their fight against terrorism. The Court does acknowledge that the fight against terrorism is very complex. It even indicates that it allows states a wide margin of appreciation in this respect. However, de facto the scrutiny by the Court is rather strict (indicating a narrow margin of appreciation).

It is within this narrow margin where human rights change and expand (or indeed contract) in response to the needs of and changes in society. In a democratic society civilians determine what their most intrinsic rights are through the law; conversely, we also live in a world where the law prescribes civilians. Apparently, the Court is careful not to allow the fears of the Zeitgeist to negate the essence of human rights protection. In other words, the elasticity of human rights is not limitless. Warnings against the proliferation of human rights and legitimacy concerns guard against over-expansion, while the margin of appreciation doctrine is kept in check so as not to entail overly drastic reductions in protection.

1 Declaration on the Right to Development 1986:
    Article 6:
    2. All human rights and fundamental freedoms are indivisible and interdependent; equal attention and urgent consideration should be given to the implementation, promotion and protection of civil, political, economic, social and cultural rights.
    3. States should take steps to eliminate obstacles to development resulting from failure to observe civil and political rights, as well as economic, social and cultural rights.
    Article 7:
    All States should promote the establishment, maintenance and strengthening of international peace and security and, to that end, should do their utmost to achieve general and complete disarmament under effective international control, as well as to ensure that the resources released by effective disarmament measures are used for comprehensive development, in particular that of the developing countries.
    Article 8:
    1. States should undertake, at the national level, all necessary measures for the realization of the right to development and shall ensure, inter alia, equality of opportunity for all in their access to basic resources, education, health services, food, housing, employment and the fair distribution of income. Effective measures should be undertaken to ensure that women have an active role in the development process. Appropriate economic and social reforms should be carried out with a view to eradicating all social injustices.
    2. States should encourage popular participation in all spheres as an important factor in development and in the full realization of all human rights.
    There is no supervisory practice so far for the rights of peoples to development and hence not a fully established right.
2 Niemietz v Germany, 16 December 1992.
3 Copland v UK, 3 April 2007.
4 Societe Colas Est v France, 16 April 2002.

SWEET REVENGE

Revenge is a recurrent theme throughout literature, art and history. Lord Byron declares, "Sweet is revenge"[1], but the poet John Milton warns, "Revenge, at first though sweet, / Bitter ere long back on itself recoils".[2] In the pursuit of happiness, one must decide: will revenge be bitter or sweet? Such decisions have consequences. In the aftermath of a recent university massacre, one woman wrote, "I don't think there would be anything temporary about the satisfaction I would feel in being permitted to execute the person who killed my child."

My colleagues and I have explored beliefs about revenge and tested the accuracy of the 'sweet revenge' hypothesis in a series of laboratory experiments. That is, do people expect revenge to feel good, and does it? To do this, we brought volunteers into our lab and created an elaborate series of interactions with other people designed to make even the most gentle and phlegmatic person seethe with anger. Some of these people were asked how they would feel if they could have got revenge against the offender, while others were actually given the opportunity to carry out that revenge. In this way we were able to learn about the emotional benefit that people predicted revenge would provide, but also the emotional benefit that revenge actually provided.

People came into our lab in small groups and learned that they would play an interactive economic decision-making game. We gave everyone some money and explained that they could keep any profit they earned from the game. The rules created a variant of the 'prisoner's dilemma': each person could do moderately well by cooperating with the others if those others, in turn, cooperated with them. Each individual, however, could make substantially more by reneging and playing competitively. But if all the players were to compete, then everybody would lose. It was, essentially, a game

1   *Don Juan*, Canto I, Stanza 124
2   *Paradise Lost*, Book IX, Line 171

of trust. People were generally inclined to be cooperative, and we encouraged this by planting a 'confederate' who exhorted everyone to cooperate. Critically, this confederate subsequently double-crossed the others in order to reap personal financial gain at the other players' expense. This manipulation, of course, infuriated the other players. How do we know this? Because the players were encouraged to text each other and the following messages (directed to our confederate, 'Dazz') were typical: "damnit dazz...YOU LIE," and "Dazz your an asshole!!!! now no body makes money... put in all your god damn money." The confederate's subsequent comments about her 'victory' exacerbated these sentiments, and when an opportunity for revenge against her was presented, over 95% of the participants took advantage of it.

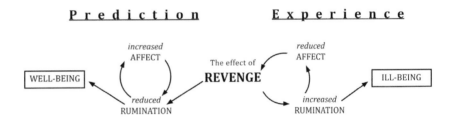

People expect revenge to set them free. They expect the act of revenge to even the scales and thus allow them to disgage from the offender. This, they think, will make them feel better. In practice, though, revenge leads to more rumination, which leads to feeling even worse. And feeling worse leads to even more rumination. This recursive relationship can lead back to revenge and create a vicious circle.

We discovered in our experiments that the revenge in people's imagination is a very different beast from the revenge in actual practice. People expect revenge to yield satisfaction and to make them feel better. The process, they think, is that the act of revenge will 'even the scales' and allow them to 'move on' in a psychological sense. They imagine that they will think less about the miscreant, and that this will lead to improved affect. To use a spatial metaphor, people seem to believe that the act of revenge will push the target away from them and away from their psychological space, and that this distance will make them feel better. The harder one pushes, the further away the other will go.

This metaphor does a good job describing people's expectations, but it does a terrible job describing what actually happens. A more accurate metaphor might describe an elastic band connecting the two individuals: the more one pushes the other via revenge, the more powerfully that person snaps back to one's psychological space. In our experiments we consistently found that people who exacted revenge felt worse than they had expected. Moreover, they felt worse than the 'control group' who experienced the same insufferable confederate but were never given the opportunity for revenge. The revenge-takers continued to ruminate about their target after the fact, and this increased rumination led to feeling worse. This dynamic created a feedback loop: the negative affect they experienced led to even more rumination, the rumination led to more negative affect, which led to more rumination, and so on.

What's most troubling is that not only do people fail to intuit this result, but also they fail to learn from experience. We asked our experimental participants at the end of their encounter about the emotional consequence of revenge. Not surprisingly our control group—who didn't have the opportunity for revenge—were quite certain that revenge would have made them feel better. But the revenge-takers—who were objectively less happy than the control group—reported that the act of revenge had made them feel better! So we have good evidence that revenge is not sweet and does not provide the anticipated satisfaction. But people tend not to believe it, even right after experiencing the downside of revenge. Indeed, keep this research in mind the next time someone slights you at a party or cuts you off on the highway; my guess is that you'll secretly believe that your revenge would be really and truly sweet.

nano-ELASTICITY ①

It began with the tip of a scanning tunneling microscope touching a silver surface. The contact being down to a single atom with a quantized conductance. We created bumps and holes on the nanoscale - the metal flowed like honey

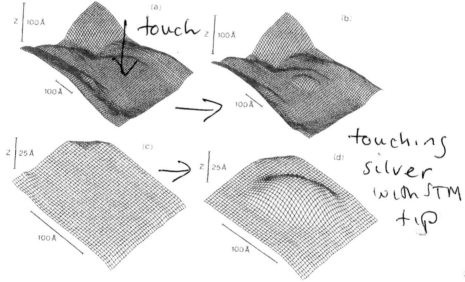

touch (a) (b)

touching silver with STM tip

Then we looked at molecules and saw they were "elastic" The legs of a porphyrin bent depending on the atomic grooves of the surface on gold and silver crystal faces the asymetry of the legs let us see the angles change from 90° to 30°

32

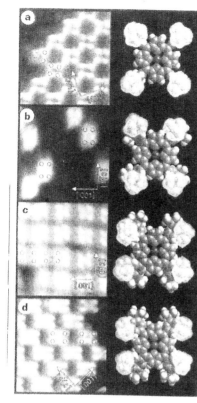

Asp. ratio ~ 1
~ 90°
Cu(100)

Asp. ratio ~ 1.4
~ 65°
Au(110) "P"

Asp. ratio ~ 1.6
~ 45°
Au(110) "F"

Asp. ratio ~ 1.8
~ 30°
Ag(110)

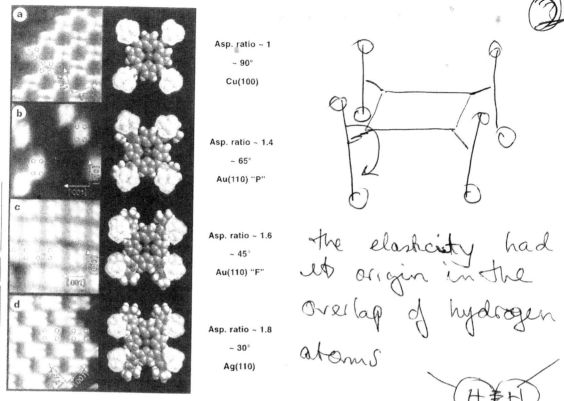

The elasticity had
its origin in the
overlap of hydrogen
atoms

H ≡ H

Then we looked at buckminsterfullerene
(C₆O) The soccer ball molecule and we squeezed
it with the tip – It came back to
its original shape. The molecular
mechanics were identical to a
soccer ball being squeezed hard

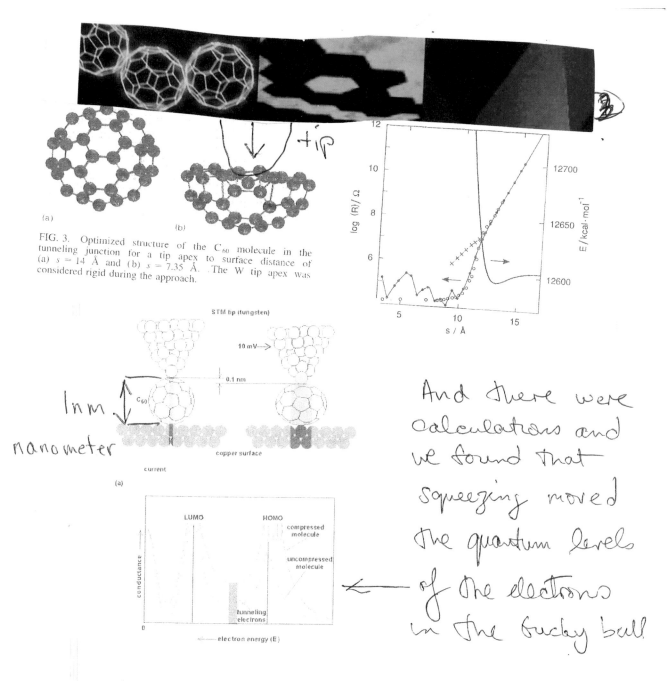

(a)

(b)

FIG. 3. Optimized structure of the C_{60} molecule in the tunneling junction for a tip apex to surface distance of (a) $s = 14$ Å and (b) $s = 7.35$ Å. The W tip apex was considered rigid during the approach.

STM tip (tungsten)

10 mV →

0.1 nm

C_{60}

1 nm
nanometer

copper surface

current

(a)

conductance

LUMO

HOMO

compressed molecule

uncompressed molecule

tunneling electrons

0

← electron energy (E)

And there were calculations and we found that squeezing moved the quantum levels ← of the electrons in the bucky ball

Well after a while we went back to the porphyrin and we looked at it with the Atomic Force Microscope (AFM) →

Which has a fine cantilever made from silicon →

hν

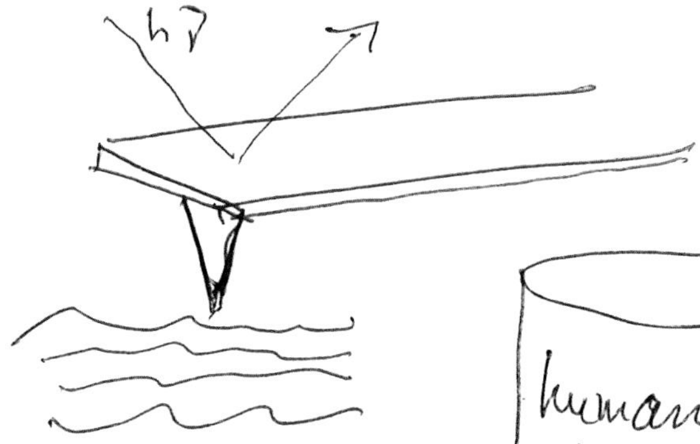

We managed to make a turning, yes an elastic one, and we recorded the force as we turned a single Carbon – Carbon bond

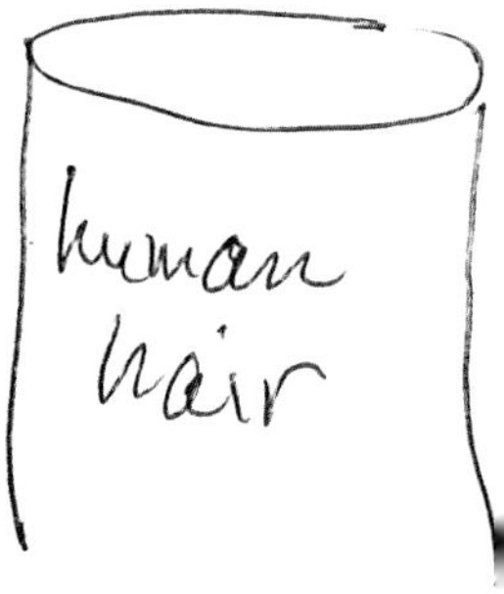

human hair

C
C

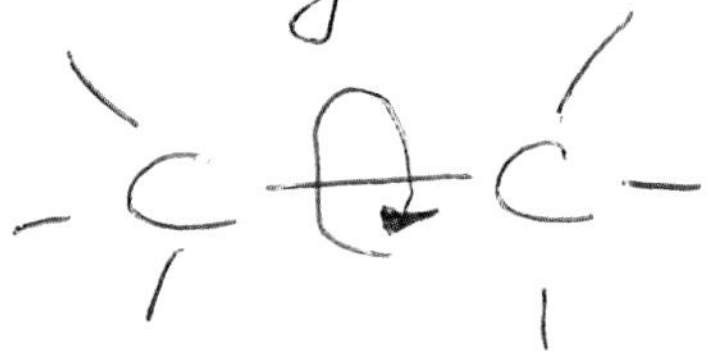

the energy was zeptojoules – close to the thermal energy of the room

VIEW LETTERS
week ending 14 FEBRUARY 2003
Force (nN)
Distance Tip–Sample (nm)
A
B
C

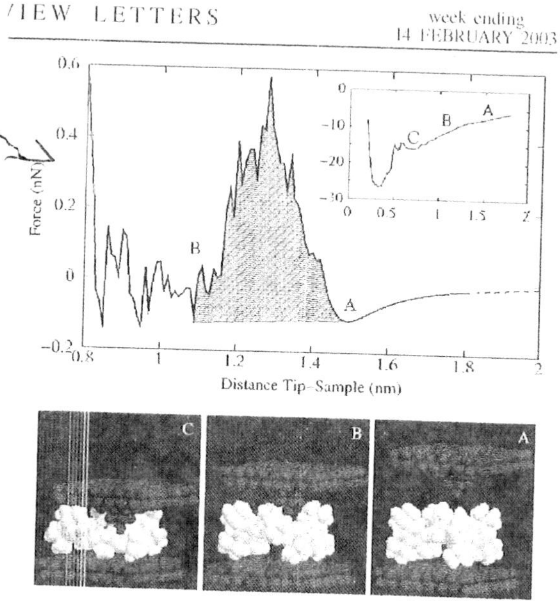

The elasticity is just electron-electron interactions reversibly going back to

Their location, What else to squeeze?

: Live cells in humans – taken from their lungs and we squeezed their cells

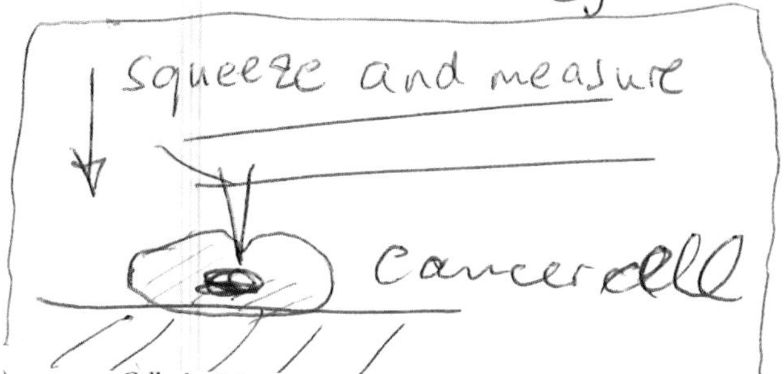

squeeze and measure

cancer cell

Cell elasticity measurements (Young's modulus, *E*) taken on cytological samples collected from patients with suspected metastatic adenocarcinoma are shown in Fig. 2. Data collected

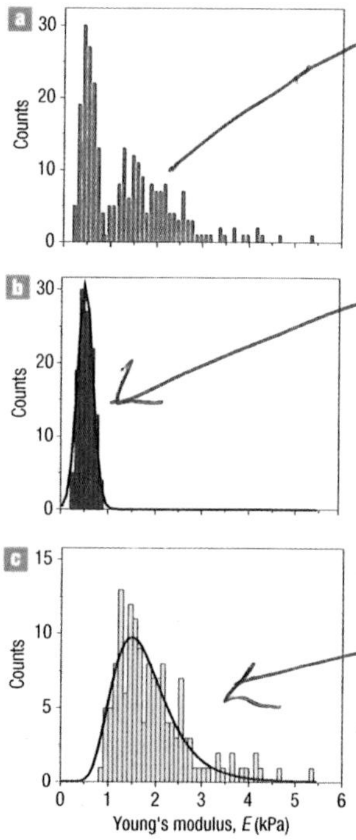

all patients

metastatic cancer cells

normal healthy cells

Young's modulus, E (kPa)

In a period of 20 years

# of cells ↑

we went from atoms

elasticity →

to molecules to cells —

...for example, refs 13–15).

In this letter we report on the associated cell stiffness (elasticity) of pathologically defined human metastatic cancer cells and benign

to seeing soft cancer cells

from a human dying

60/Female	Non-small cell carcinoma of the
49/Female	Breast ductal adenocarcinoma
85/Male	Pancreatic adenocarcinoma
40/Male	Liver cirrhosis
7/Male	Fever and hepatic failure
2/Female	Anasarca peripheral oedema

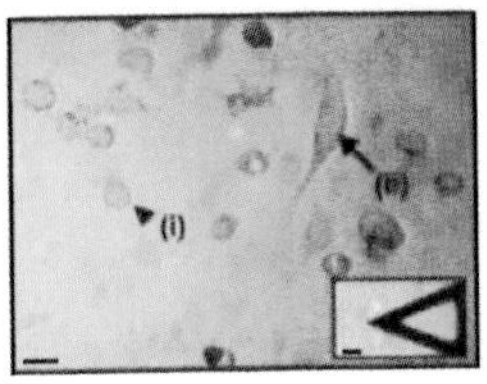

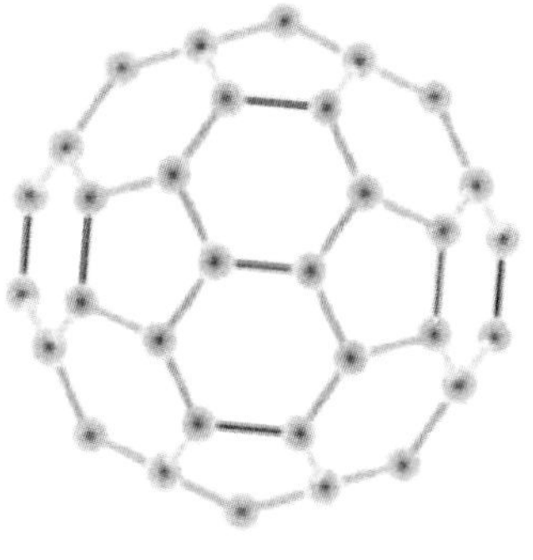

Buckyball C60 sphere

⌐DNA ⌐

but theres more — we streched DNA — more elasticity, we looked at it and then "bit out" AGCT sections and looked a the breaks where it jumped back. We took those concepts and created Art/Sc where metaphor of STM (became) a human hand

Here was Something new Allegra Fuller danced and squeezed virtual buckyballs at the Los Angeles County museum of Art

Art/Sci is the elastic interplay of two brain lobes touching!

James Gimzewski 2009

Allegra Fuller, Sir Buckminister's daughter, reaches for a Buckyball.

Scanning Tunneling Microscope designed to manipulate
molecules at ultra low temparatures.

With Reto we built the microscope
we touched many things, squeezed
prodded — it is a journey to
nanospace — a beautiful one

I share the experience through
science, art, talking, doing

THE ENTROPY OF DNA

Elasticity is an essential factor for understanding the physics of biological material. I am confronted in my work with elasticity at very small length scales (10^{-9} m i.e. the *nanoscale*). The elastic properties of these materials reveal important information about their biological role and function. In this piece I'll discuss entropic elasticity of (long) bio-polymers and the information it can reveal of protein binding to these bio-polymers.

BIO-POLYMERS

Cells employ an eclectic array of bio-polymers to perform a diverse set of tasks: from actin and intermediate filaments, which provide cellular structure; and microtubules, which form the intercellular transport network; to DNA, which is the molecular basis of the genetic code. Each of these bio-polymers is regulated and structured by proteins that remodel these filamentous networks according to the needs of the cell.

PHYSICAL PARAMETERS

To physically describe bio-polymers we need a number of essential parameters. The most important ones are *length* and *rigidity* (or *persistence length*). The property *length* (L) is clear. The *persistence length* (P) is less intuitive; it describes how stiff the polymer is. The mathematical expression of P is:

$$\exp\left(-\frac{s}{P}\right) = \left\langle \cos\theta(s) \right\rangle \text{ with } \left\langle \cos(\theta_{ij}) \right\rangle \alpha \left\langle \vec{x_i} \bullet \vec{x_j} \right\rangle$$

where s is the distance between two vectors (x_i & x_j) along the polymer, θ_s is the angle between the vectors separated by distance s. In non-mathematical terms the persistence length can be described as the distance between two locations along a polymer that can be reoriented freely. In other words if you have a polymer which is a persistence length long, then you can point the two ends in a random orientation with minimal effort. A floppy material such as a thin rubber band has a small persistence length and a stiff object like a straw has a long persistence length.

I typically work with DNA, either double-stranded DNA (dsDNA) or single-stranded DNA (ssDNA). The persistence length of dsDNA is 45 nm and ssDNA 0.5 nm. These are very short compared to the typical length of a DNA molecule (10-1000 μm). Changing the shape of an object which is much longer than its persistence length takes very little energy. DNA is usually dissolved in liquid and therefore constantly bombarded by molecules. The energy supplied by these collisions constantly shakes the DNA into many different (globular) shapes (or states). The number of states a system can take represents its entropy. The 2nd law of thermodynamics postulates that a system strives to increase entropy. At the entropic maximum the DNA molecule fits inside a sphere with radius:

$$R_g = \frac{P}{3} N^{1/2} \text{ with } N = \frac{L}{P}$$

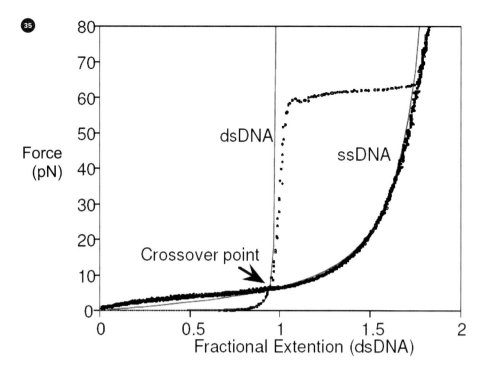

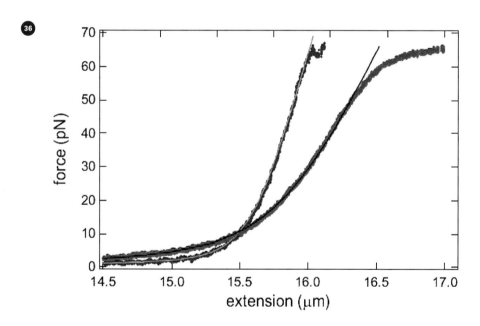

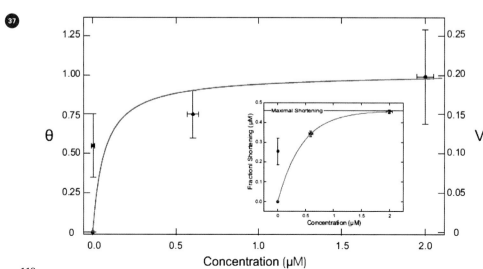

118

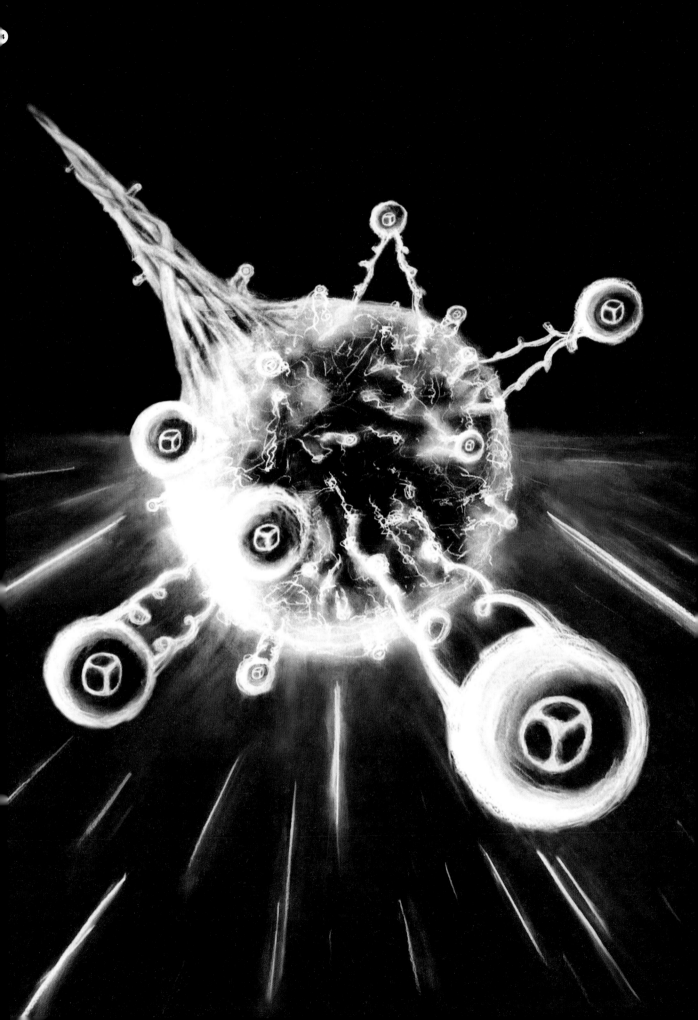

If the two ends of a DNA molecule are grabbed and the DNA is forced in an extended shape it loses the possibility to be in so many different states. This effect is counter to the physical drive of the 2nd law of thermodynamics and as a result it takes force to put a DNA molecule in an extended state. *This effect is the essence of entropic elasticity*. The entropic elasticity for dsDNA molecules is given by:

$$F = \frac{k_b T}{P} \left(\frac{1}{4 \left(1 - \frac{x}{L}\right)^2} + \frac{x}{L} - \frac{1}{4} \right)$$

where k_b is the Boltzmann constant, T the temperature and x the distance between the ends of the molecule. Figure 35 displays this relation (red) for a dsDNA molecule of 16 μm and a ssDNA molecule of 27 μm (but they have the same number of base pairs).

(35) Force–extension relation of dsDNA and ssDNA. In black, actual measurements of the entropic elasticity of dsDNA and ssDNA. In red, the result of the mathematical description for this entropic elasticity.

PROTEIN BINDING TO DNA
Many proteins deform DNA when binding. This involves stretching, bending and/or kinking and unwinding of the double helix. The energy for the distortion is usually provided by multiple favourable contacts between DNA and protein. DNA binds to a special region of a protein. This binding site has the form of a cleft and is usually non-polar. The binding site takes up a rather small portion of the total protein. The remaining amino acids act as a scaffold to form the right shape for substrate binding. Substrate binding is governed by many weak non-covalent bonds, all in the order of $k_b T$. This is of crucial importance, because proteins must bind reversibly to their substrate. The interactions involved are electrostatic interactions, hydrogen bonds and Van der Waals interactions. The equilibrium constants of enzyme-substrate complexes typically range from 10^{-2} to 10^{-8} M. This corresponds to energies ranging from 5 to 20 $k_b T$. The weak Van der Waals interactions only play a significant role on very short distances. To take advantage of this, the substrate and binding site must have complementary shapes in the complex. Hydrogen bonds on the other hand also require dipoles somewhat aligned to each other. These restrictions mean that proteins are highly specific for particular substrates. Due to binding, many proteins change the physical parameters such as the *entropic elasticity* of DNA. We can measure these before and after protein binding. The changes in these parameters give us detailed information about the protein binding. Typically, this information can be used to elucidate protein function, which can have broad implications in, for instance, diagnostics of cancer-causing mutations in protein (see under DNA-stretching experiments).

ALBA
Currently I am investigating ALBA, a DNA binding protein from the Sac10b family that is highly conserved throughout the archaeal domain (single-celled microorganisms). Alba makes up 4–5% of the total soluble protein of most archaea. ChIP experiments have demonstrated that Alba is distributed uniformly on the chromosome; there is no evidence that Alba binds DNA sequence specifically.

Biochemical studies have not yet provided much understanding of Alba-DNA interactions at the structural level. Moreover, single-molecule studies that have been done so far give limited insight into the way in which Alba interacts with DNA. In this study we have combined several single-molecule techniques in order to gain more insight into the molecular-mechanism underlying Alba-DNA interaction. Taking two proposed models as starting point I try to characterize how Alba binds to one or two DNA duplexes. I aim to describe this interaction in a structural as well as a thermodynamic way. Furthermore I am interested in how our findings translate into *in vivo* functions that Alba might have.

DNA STRETCHING EXPERIMENTS
We have performed DNA stretching experiments in buffers with an Alba concentration from 0 to 2 μM. This enables us to observe the influence of Alba on a single DNA molecule and probe the kinetics of the Alba-DNA interaction.

(36) Force-extension curves of single DNA molecules fitted with the WLC model; red: bare DNA, blue: DNA in 2 μM Alba.

As can be seen in figure 36, a dsDNA molecule stretched in a buffer containing 2 μM Alba shows a different force response than a molecule stretched in the same buffer lacking Alba. The most striking feature of this trace is that the slope is a lot steeper in respect to a force-extension curve taken in a buffer without Alba. Intuitively, it is easy to couple this steeper response to a stiffer DNA molecule. When this trace is fitted with the WLC model, indeed a longer effective persistence length is found. Additionally, fitting also yields a contour length that is decreased compared to that of bare DNA. By performing these experiments at different Alba concentrations we found both effects to be concentration dependent (see table). At saturating Alba concentrations DNA stiffens ~3 fold and shortens ~3%.

Alba concentration (nM)	P (nM)	L (μM)	N
0	38.8 ± 1.45	16.49 ± 0.026	41
2	64.9 ± 6.29	16.23 ± 0.068	11
600	147 ± 5.61	16.14 ± 0.014	4
2000	136 ± 8.48	16.03 ± 0.010	26

Persistence length and contour length of DNA. N represents the number of trace taken at the specified concentration. Errors are SEM.

MCGHEE VON HIPPEL ANALYSIS
The influence of Alba on the persistence length and contour length of DNA is not the only information that we can extract from optical tweezers experiments of DNA. McGhee and Von Hippel have developed a theoretical framework that describes the kinetics of ligands binding to a polymer; we use this theory to characterize the binding of Alba to DNA. The McGhee-von Hippel binding isotherm describes the concen-

tration dependence of the binding density ν of the ligand with two thermodynamic parameters: ligand site size n and intrinsic binding constant K, where C is the concentration of the ligand.

$$\nu = KC\,(1 - n\nu)\left(\frac{1 - n\nu}{1 - (n-1)\nu}\right)^{(n-1)}$$

The data in the table show that the Alba-induced shortening of DNA is concentration dependent. Therefore, we can use the shortening as indicator for the binding density. To determine the fractional shortening of DNA at a certain concentration Alba, we have to make an estimate of the maximal shortening possible, i.e. the shortening that would result if a binding density of 100% would be reached. As can be seen from the inset in figure 37 the Alba induced shortening of DNA saturates. Fitted with an exponential function this yields the Alba induced shortening that would result at full DNA saturation.

McGhee-von Hippel analysis. Titration curve of fractional shortening data fitted with the cooperative McGhee-von Hippel binding isotherm. Inset: evolution of fractional shortening with concentration.

McGhee and Von Hippel pointed out that, for n > 1, the maximal binding density obtained experimentally does not necessarily correspond to a binding density of 100%. This stems from the fact that free spaces smaller then n can form between bound ligands. Because of this, a molecule may appear saturated without having a binding density of 100%. The DNA saturation θ, is related to the binding density via the binding site size: $\nu = \theta/n$. We can now calculate the fractional shortening as:

$$\frac{L_0 - L_C}{L_0 - L_{max}} = \theta = n\nu$$

in which L_x corresponds to the contour length of DNA at concentration x and L_{max} to the estimated maximal contour length at saturating Alba concentration. From various experiments the footprint of Alba has been estimated to be ~5 bp. Using this value we are able to fit the concentration dependent binding density. Since we use the footprint to calculate ν, we fix n to 5 while fitting. The binding constant $K = 3.0*10^6 \pm 3*10^5$ M^{-1} we obtain seems quite high when compared to the binding constant of other ligands obtained using McGhee-von Hippel theory. The binding constants of vaccinia topoisomerase IB to dsDNA, ethidium to dsDNA, and T4 UvsX Recombinase to ssDNA were previously measured. All three ligands were found to have a K in the order of 10^4 to 10^5 M^{-1}, roughly 10% of the K we obtained for Alba. We can conclude from the McGhee-von Hippel analysis that Alba readily binds DNA.

DISCUSSION

We have observed that at increasing concentrations DNA gets coated with Alba. We suggest that this results in the formation of an Alba filament on the DNA. When Alba binds DNA, the bound proteins will be rotated relative to each other following the helical structure of the DNA; this will thus result in a helical filament. This filament might slightly bend the DNA it wraps which could be responsible for the shortening we observe. The filaments that Alba might form on DNA when present in high concentrations may render DNA inaccessible for other proteins.

These conclusions about the functional role of ALBA have been obtained by the study of the physical properties of the DNA protein filament. This step could only be taken because we have a solid understanding of the elastic and mechanical properties of DNA without proteins bound to it. Essential in this understanding is the realization that the entropy of DNA molecules forms the basis of its elastic behaviour.

Sources:
Dissertation: Maarten Noom, *Mechanisms of DNA organization unraveled with novel single-molecule methods*. Under supervision of G.J.L. Wuite, 06/2008
Dissertation: Bram van den Broek, *Cleaving, condensing and manipulating single DNA molecules*. Under supervision of G.J.L. Wuite, 06/2007
Master Thesis: Felix Hol, *Architectural properties of the archaeal DNA organizing protein Alba*. Under supervision of G.J.L. Wuite, 02/2008

ELASTIC BALL

The world's largest rubber band ball was created by Joel Waul, first in his home and then driveway in Lauderhill, Florida. It weighs 4096.84 kg and is more than two metres tall.

It took more than 700,000 rubber bands and set the world record on November 13, 2008.

LONDON – Hollywood star Bruce Willis has invented a bizarre helmet to help avoid head injuries on the sets of his action films. The 54-year-old gave American TV audiences a laugh when he banged and bounced his head on the desk of talk show host David Letterman May 6 night to prove how well the device works, reports femalefirst.co.uk. "I'm feeling a little clumsy lately. A little clumsy around the house and clumsy on the sets. You know, you get a little kicked in the head sometimes, banged around. I'm getting a little older so I was fooling around the house and I get one of those little rubber band balls. This protects my head. I like to call it Bruce Willis' Concussion Buster," he said. Laughs also turned to gasps when Willis challenged the host to smash an empty bottle of whiskey against his bald head after he took off the gear, sending glass shattering all over the set. Willis shook off the stunt, joking: "It's the seatbelt of the 21st century."

Rubber

http://blog.taragana.com/e/2009/05/10/bruce-willis-invents-bizarre-headgear-3642/

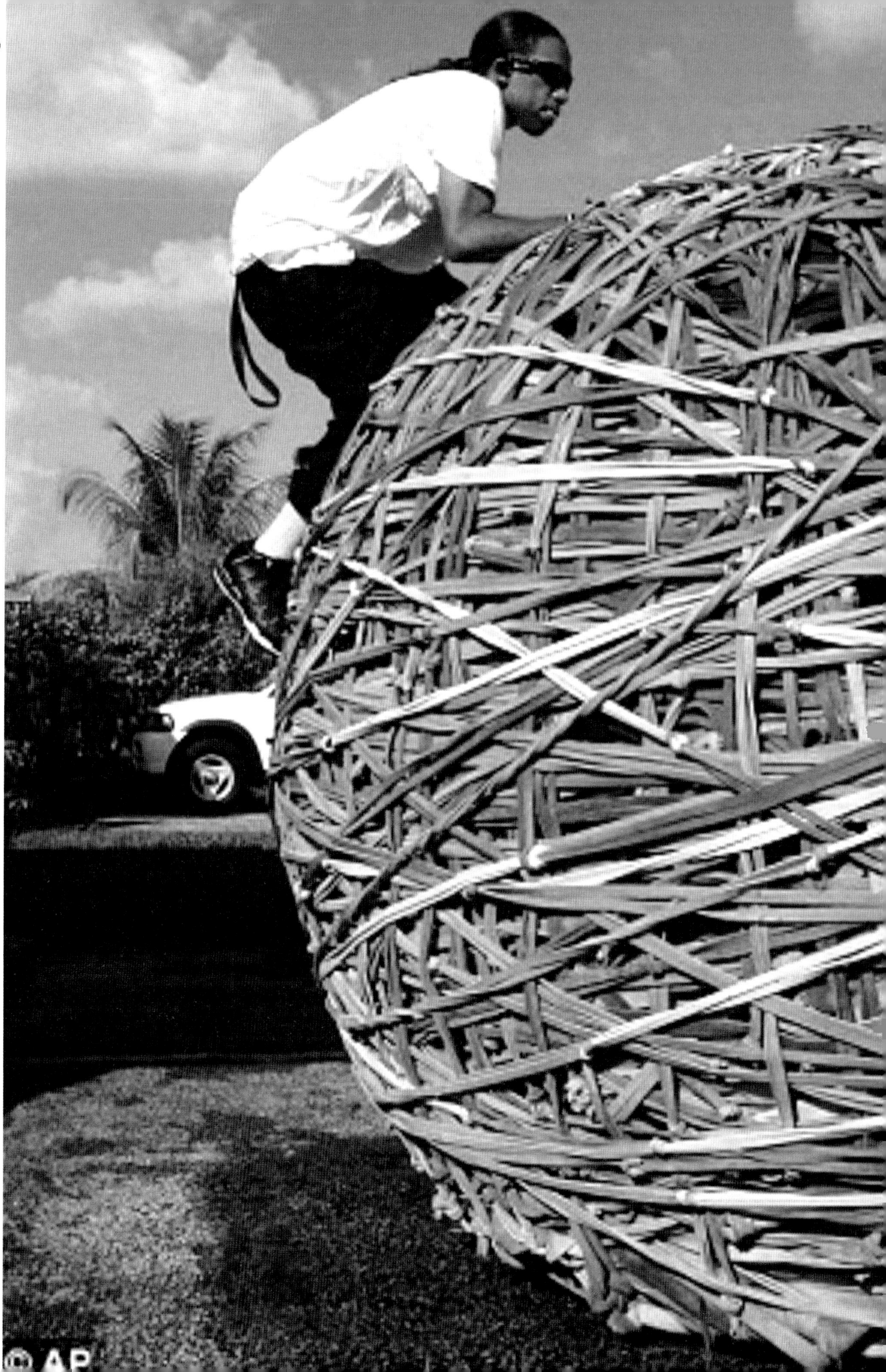

A MATERIAL OF CHOICE

Take a deep breath. Hold it. Now breathe out as much as you can. Repeat 10 times.

That is the total amount of air an average condom will take before it bursts.

The rubber film in a condom starts out about 60 to 70 microns thick—just over half of one tenth of a millimetre. The condom is about 18cm long and the circumference is just over 10cm.

If you sit down and do the arithmetic, that means that just before the condom bursts at about 40 litres, the rubber film has gone from being over 60 microns thick to about 1 micron thick—that's 17 times thinner than the thinnest human hair.

The air burst test is used in laboratories worldwide as a measure of the overall strength of a condom. You record both the burst volume and the burst pressure and there are minimum allowed values for each.

39 Several years ago researcher Andrew Davidhazy decided to try and take pictures of the condom at the moment of burst. Using very fancy high speed photography, special lighting, and noise detectors for the first appearance of the bang, he found the condom body usually splits open with a crack rapidly forming up and down the length. In the twinkling of an eye, the crack propagates up nearly to the closed end of the big inflated balloon and down to the place where the condom is clamped for the test. At each end the crack then usually takes a 90° turn and races round the circumference to join up with itself. At that point the burst is over and you are left with three pieces. From the closed end is a hemispherical cap with the teat. From the middle is a flat sheet, and from the open clamped end is a ring. The energy that was stored in the stretched condom due to elasticity is converted into a puff of air and a loud bang.

Of course, it isn't always exactly like that. So much energy is stored in such a small volume of actual rubber (only about one third of a teaspoonful) spread over such a large area that it isn't always released exactly the same way. It's a bit like trying to smash milk bottles with a hammer in a repeatable way. No matter how much control you put into the hammer, the results always vary a bit.

Most condoms are made from natural rubber which comes from the sap of a particular tree that originated in the Brazilian rainforest before being cultivated. Now the tree is grown mostly in the Far East in plantations.

Today we also have a synthetic version made from petrochemicals in big chemical reactors.

The chemical name for natural rubber is cis-polyisoprene. The isoprene is the chemical, the poly means lots of isoprenes are joined up end to end and the cis means they are joined up in a particular way.

Nature, naturally, is brilliant at joining all the isoprenes up in this particular way in this particular tree. More recently the chemical engineers have managed it too. Surprisingly (well, I was surprised), condoms made from synthetic cis-polyisoprene are even more elastic and have an even bigger burst volume than natural rubber condoms.

A natural rubber condom can average 40 litres at burst. Synthetic polyisoprene can average up to 70 litres and a few individual condoms make it over 100 litres. Highly impressive though this is, it proved to be a bit of a nuisance in the test laboratories because, firstly, the standard soundproof burst cabinets were too small and, secondly, since we use a constant air flow rate, the test takes a lot longer to complete.

Because of the condom's shape, when it is inflated it stretches more width-wise than it does length-wise. You can tell, like an Australian researcher did, by drawing a square on the un-inflated condom and watching the shape change to an oblong as the condom inflates.

For natural rubber, the widthwise stretch is about six times and lengthwise about four times. Materials scientists call this 'elongation' of 500% and 300% respectively. The shape of the condom means that on inflation there is more stress around the circumference than there is along the axis. So the stretch is more around the circumference. When the film eventually fails, the split is 90° to the main stress and therefore in the same direction of the condom axis.

Interestingly, if you take a ring cut from a condom using a pair of parallel blades and stretch it in one direction, the elongation at breakpoint is about 800%. In an inflation test the stretch is in two directions and the combined area elongation is the product of the two stretches and much higher.

Some materials are much stronger in one direction than another, string for example. Rubber doesn't seem to mind and the reason usually given for this is because it's actually rather like a liquid with no preferred direction for strength.

Coming out of the tree, natural rubber is in the form of tiny rubber spheres a few microns in diameter separated by a water-based serum. The serum plus the dispersed rubber spheres is called latex. If the water is allowed to evaporate, the spheres start to touch each other and they stick together. If all the water evaporates you end up with a lump or film of rubber and, short of smashing it up in a very aggressive way at very low temperatures, you can't get back to the original particles again.

The tiny rubber spheres stick together for two reasons. Firstly, the chain molecules of cis-polyisoprene are very flexible because the individual chemical bonds in the backbone are able to rotate very freely. Think of it as a well lubricated chain. This means that down to about -70°C there is enough heat energy constantly vibrating the rubber molecules to make them mobile—again a bit like a liquid. If two tiny spheres of rubber touch by accident, the flexible chains have enough mobility to intermingle a bit, so the particles stick.

Secondly, the rubber itself (polyisoprene) is a hydrocarbon. Like other hydrocarbons it has low surface energy. The tiny rubber particles are surrounded by water which has higher surface tension and so the two don't mix and want to separate. Oil and water, in fact.

So when the water evaporates from the natural rubber latex and the rubber particles start to touch, they will stick together because they are alike in surface tension. They 'wet' each other. The resulting reduction in surface area exposed to the water means there is a reduction in surface energy and so there is a 'driving force' to keep them stuck. This is like two droplets of oil on the surface of a puddle coalescing. Finally the intermingling of the rubber chains by diffusion means that the original identity of the spherical particles at the surface is lost at the surfaces in contact.

In a condom, the latex particles coalesce to form a solid in the form of a film. If we stretch the film in one direction each original sphere would form an elongated rod. If the stretch was in two directions then each sphere would form a plate.

Now in one respect rubber is not like a liquid in that it retains its shape and doesn't take up the shape of the vessel it is in. Despite the many ways in which rubber is like a liquid (chain mobility and so on) the one difference is elasticity. This is the tendency to recover the original shape. It takes energy to distort the shape of the rubber and nearly all that energy is recovered when the rubber returns to its original shape.

Straight out of the tree, natural rubber has a marked tendency to creep. If you stretch it slowly and don't let go, the rubber will flow like a viscous liquid to the new shape and be pretty weak. Only if you stretch it and let go very quickly will it snap back. It's pretty easy to see how creep makes untreated rubber unsuitable for applications such as condoms, tyres or shoe soles.

Synthetic polyisoprene is exactly the same, fairly useless straight out of the reactor.

The process that stops rubber creeping is called vulcanisation. Instead of allowing the long cis-polyisoprene chains to flow past each other, they are tacked together by chemical bonds every now and again, usually using sulphur. This tacking (or cross linking) allows the elasticity to be retained, but gets rid of the creep and you get a highly useful material. The trick for

39

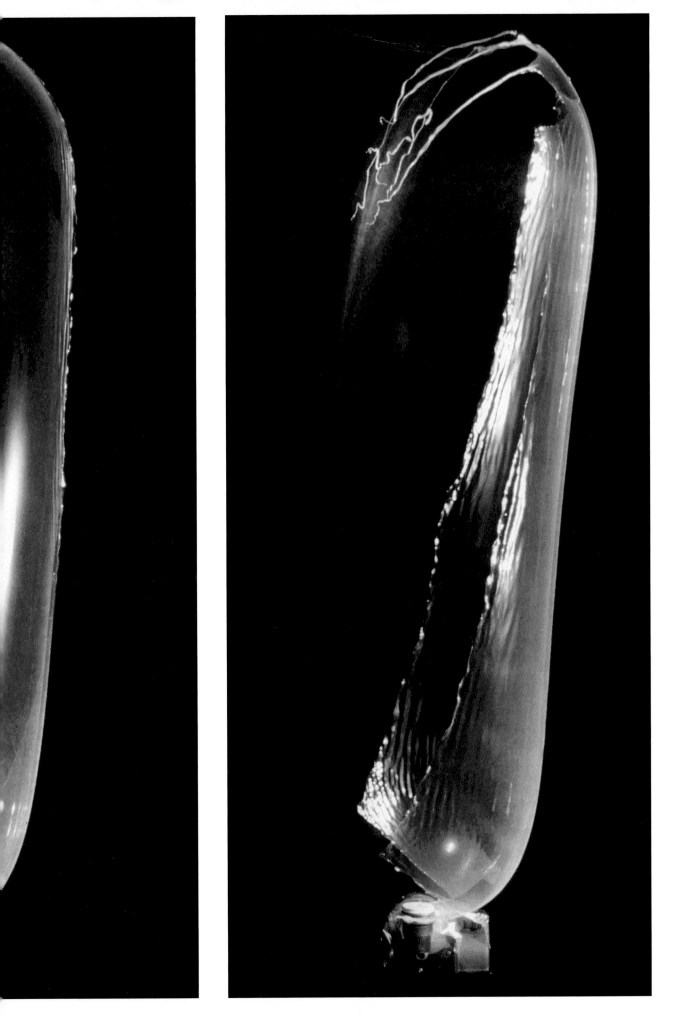

the rubber formulators is to add just enough sulphur to crosslink rubber by just the right amount for the application. Too little and you still get weakness, creep and instability. Too much and you lose elasticity and eventually get a stiff, brittle material.

Something interesting happens when you crosslink rubber and then stretch it. The chains flow over each other (slightly restricted by the cross links) and they begin to line up. As they line up, they begin to fit together because they are all the same molecule, and a temporary crystalline structure begins to form. The temporary crystallinity makes the rubber more like a solid and less like a liquid and the stiffness of the rubber begins to rise. Because the stiffness rises, so does the strength—strength is the amount of force or energy needed to break the material. So rubber is a lot stronger than you would expect because of stress crystallisation.

Strength is usually measured as the force per unit area required to break the material. In metric SI units a typical figure for rubber might be 25 mega Pascals, the units named after the French scientist. This is quite a low figure compared to other materials like, say, steel. However, the unit area used in the calculation of rubber strength is of the original rubber material and if it stretches by 800%, then the area of the sample tested decreases by exactly this much. So at break the 'true' strength is more like $25 \times 8 = 200$ mega Pascals, a much more respectable figure more similar to steel.

You can experience stress crystallisation yourself easily enough. If you take an ordinary rubber band, when it first stretches it only takes a small amount of force, but as the stretch increases the stiffness gets greater and greater until just before it breaks it feels like you have a very different material than the soft band you started with.

In case you hadn't realised, if you take a 1cm cube of rubber and quadruple its length, the width and depth reduce by half. The stretched block has a different shape to the original cube, but it has almost exactly the same volume. The change of volume on stretching is called the Poisson ratio after another French scientist. For rubber this ratio is almost exactly 1, meaning that although you can easily distort rubber into a different shape, you can't actually compress it to a smaller volume. If you put a tight-fitting circular plug of rubber into the barrel of a syringe, you would find you couldn't use the plunger to compress it because it would be exceedingly stiff.

So what is it that makes rubber return to its original shape? What causes rubber elasticity?

Everything in the universe is heading towards a lower energy state. If you have a ball at the top of a slope it will run down to the bottom. Teenagers prefer to be horizontal to being vertical. This is one of the laws of thermodynamics. Now there are two sorts of energy. The first is regular or normal energy called 'enthalpy'. Heat from a burning candle is enthalpy. The other sort of energy is called 'entropy'. This is a sort of randomness factor. A more random arrangement of something has higher entropy, a less random arrangement, lower entropy. Another law of thermodynamics is that entropy in the universe tends to increase, so the universe is gradually getting more random.

Human activity often appears to buck this trend in that we organise things—but it is always at the expense of more randomness elsewhere. For example we may organise a small bit of the universe by growing a field of corn, but this is at the expense of consuming a fair amount of energy to do so. All that energy consumption entails taking organised molecules like hydrocarbons and changing them into less organised molecules like carbon dioxide. So in creating a field of corn, we still make more randomness than we reduce. Entropy overall goes up.

A lump of rubber contains rubber chains orientated every which way like a saucepan of cooked spaghetti that has been thoroughly mixed. This arrangement has a certain amount of randomness or entropy. If you stretch the lump of rubber, the chains start to align themselves in the direction of the stretch and the more you stretch the more aligned they become. This alignment is a more organised, less random arrangement for the rubber molecule chains, so the entropy *decreases* and it takes energy to get them more organised. Because the rubber chains are very mobile, it takes very little enthalpy energy to distort them. In that sense the rubber is again like liquid.

So, because the entropy decreases as the rubber is stretched, it takes energy to stretch the rubber. But since the chains are locked together by the sulphur cross links and the rubber doesn't creep, they can't stay where they are put. They want to go back and this releases the gained entropy energy. This is a very technical way of saying the rubber is elastic. You have to put energy into the rubber to stretch it but then you get that energy back when you release it. Elastic.

The *useful* bit about the rubber is that because the chains are long, very flexible and only cross-linked quite far apart, you can distort or stretch the rubber a lot (800%, remember, for natural rubber) and it will still return. The other useful bit is that energy changes due to entropy changes are typically quite small compared to enthalpy changes, so it doesn't take much energy to distort or stretch rubber. It is soft. Soft and very stretchy equals useful.

So there you have the reasons why rubber is elastic and why it is a useful material. These properties are why it is still the material of choice for condoms.

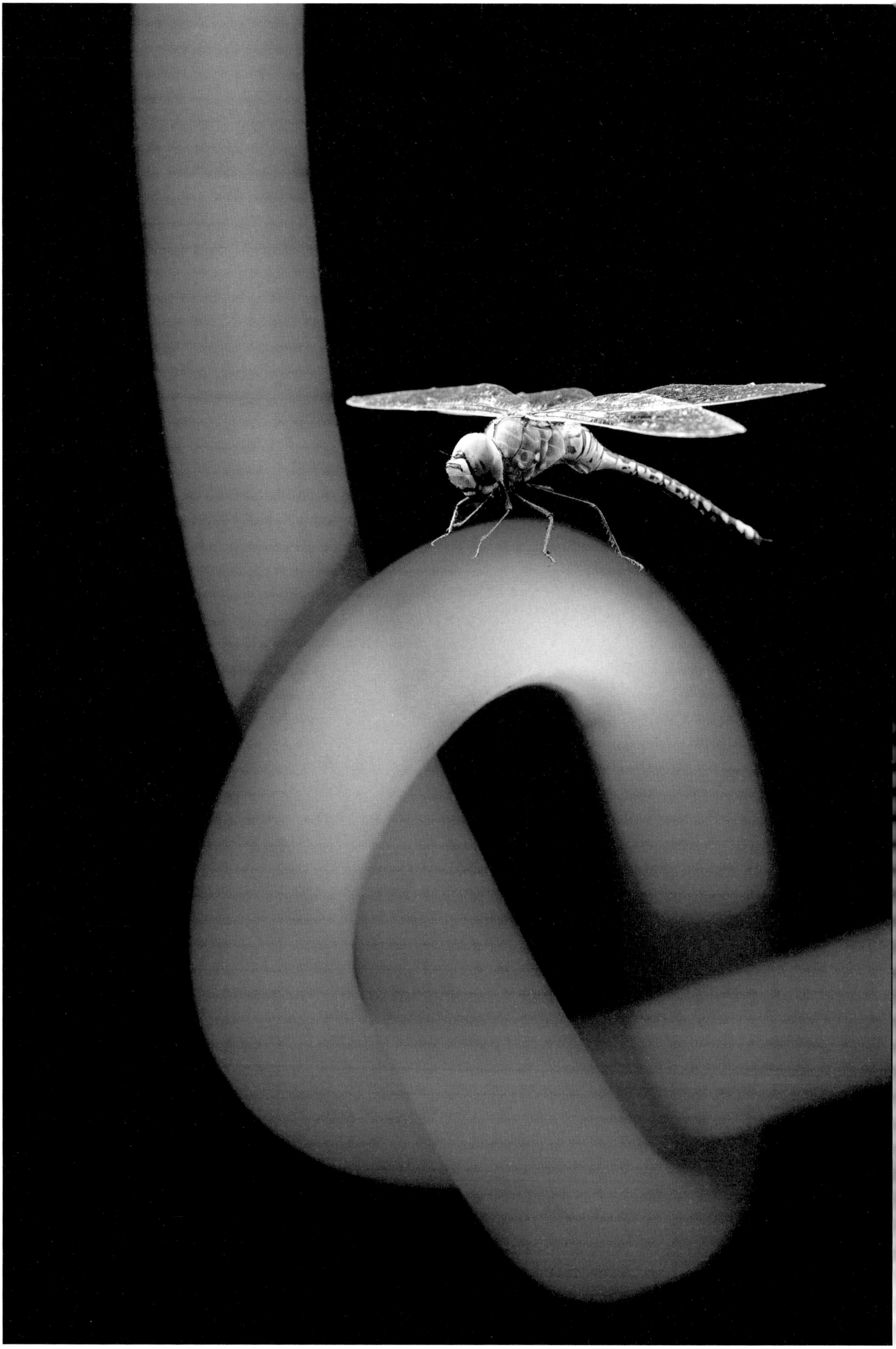

41

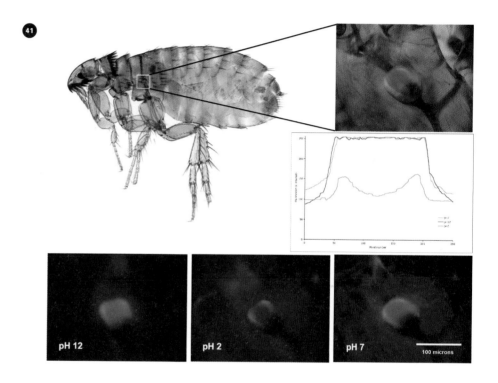

pH 12   pH 2   pH 7   100 microns

42

a

2 tyrosine $\longrightarrow$ dityrosine + 2H$^+$ + 2e$^-$

d

soluble pro-resilin

crosslinked resilin

tyrosine

dityrosine

Retention time (min)

b

resilin   tendon

0.1 mm

c

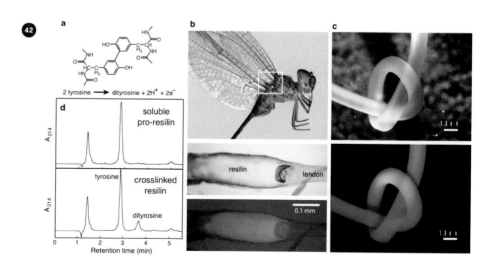

RESILIN—A NEAR–PERFECT ELASTOMERIC PROTEIN

40 Dragonfly 'perched' on a UV-illuminated rod of crosslinked recombinant resilin. Resilin knot photograph by Dr David Merritt (UQ, Brisbane, Australia);

Natural rubber and its remarkable properties have been known since the sixth century, as evidenced by an Aztec picture of a priest sacrificing two balls of rubber. Even Christopher Columbus, after his second voyage to the New World in 1496, told stories of the natives in Haiti playing with a ball that bounced very well that they made from the gum of the tree *Hevea brasiliensis*. The South Americans called the material 'caoutchouc' meaning weeping wood, a name that is still used in many languages today. In the 1770s the English scientist Joseph Priestley coined the term 'rubber' because of the material's ability to rub out pencil marks and this name has stuck in spite of its limited use for this purpose. In 1826 Michael Faraday deduced the chemical formula of natural rubber as C_5H_8. However, in spite of physical evidence from freezing point depression and osmotic pressure experiments the material was largely believed to be an aggregate of small cyclic molecules rather than a long chain macromolecule or polymer.

It was not until 1920 that Hermann Staudinger firmly proposed long chain macromolecular structures for rubber as well as polystyrene and poly(oxymethylene). He championed his idea of large macromolecules for a further 10 years before they became commonly accepted and eventually he was recognised for his pioneering achievements as the forefather of polymer science. Somewhat belatedly, he was awarded the Nobel Prize for Chemistry in 1955 at the age of 72.

We now know that many of the unique properties of polymers are largely due to the long chain-like structure of the molecules. Elasticity is a prime example. Polymer molecules, particularly amorphous ones with sufficient internal mobility, are normally in a randomly coiled structure. When such a molecule is stretched, the number of conformations (possible spatial arrangements) of the molecule decreases and hence the entropy decreases, leading to a retractive force on the ends of the molecule. This force increases at higher temperatures leading to the unusual, so-called thermoelastic effect first observed by Gough in 1805, namely that a stretched elastomer gets shorter when heated. This effect is summed up in the thermodynamic equation for an ideal elastomer:

$$F = -T(dS/dL)_{V,T}$$

where $(dS/dL)_{V,T}$ is the change in entropy (S) with a change in length (L), keeping volume (V) and temperature (T) constant.

The outcome of this thermodynamic approach was developed by Wall (1942), James and Guth (1943) and Flory and Rehner (1943) into the statistical theory of rubber-like elasticity with the widely used equation:

$$\sigma = NRT\ \upsilon^{\frac{1}{3}} * \left(1 - \frac{2M_c}{M}\right) * \left(\alpha - \frac{1}{\alpha^2}\right)$$

where σ is the stress (force per swollen, unstressed cross-sectional area), N the crosslink density (equal to ρ/M_c), R the gas constant, T the absolute temperature, υ the volume fraction of rubber in the swollen sample, ρ the unswollen density, M_c the molecular mass between crosslinks, M the primary molecular mass and α the extension ratio. Paul Flory won the 1974 Nobel Laureate in Chemistry for his fundamental achievements, both theoretical and experimental, in the physical chemistry of macromolecules.

While natural rubber in plants was playing an important part in the development of polymer science and an understanding of elasticity, insects had been quietly getting on with evolving a remarkable elastomeric protein, resilin, for their own function and development.

Elastomeric proteins are a diverse group of structural proteins present in widely divergent organisms from insects and molluscs to higher mammals. They are structural proteins displaying reversible deformation and serve an important functional role in providing long-range elasticity, storing kinetic energy, acting as shock absorbers and as antagonists to muscles. The proteins possess rubber-like elasticity undergoing high deformation without rupture and then returning to their original state on removal of the stress, passively releasing energy stored on deformation. Such an entropic mechanism is ideal for elastomers that are required to last a long time. The family of elastic proteins includes the vertebrate muscle and connective tissue proteins, elastin, titin, and fibrillin, as well as the seed storage proteins gluten and gliadin, the bivalve ligament protein abductin, mussel shell byssal threads, spider silks, and the rubbery protein resilin, from arthropods.

Despite diverse protein sequences among the members of this protein family, all elastomeric proteins share a number of common features. Firstly, they all contain repeat peptide motifs comprising both elastic domains, which are rich in the amino acid glycine and other hydrophobic amino acids, as well as nonelastic or 'cross-linking domains', involved in the formation of obligatory chemical cross-links in the mature polymer. Secondly, they always occur in nature in a hydrated state, the water acting as both lubricant and plasticizer, and resulting in structures that are highly mobile and conformationally free. Finally, all of the proteins are covalently cross-linked, although the chemical nature of these cross-links varies. In natural resilin from insect joints and tendons, crosslinking occurs between tyrosine residues, generating diand trityrosine. Analysis of photochemically crosslinked recombinant resilin reveals dityrosine crosslinks, reaching similar levels to that found in the resilin isolated from native insect joints. The fact that dityrosine fluoresces bright blue under UV light allows us to visualise quite dramatically the presence of dityrosine in both the natural and recombinant materials (Figures 40, 41 and 42). The structure of dityrosine is shown in Figure 42a.

Fleas can leap up to 35 centimetres in a single bound—about 200 times the length of their own bodies. This would be equal to a 300-metre jump by a two metre high human. Bennet-Clark and Lucey showed in 1967 that resilin was the spring that powered the high-speed catapult used in flea jumping (Figure 41). Spittle bugs experience a force about 400 times that of gravity during a jump and they accelerate three times faster than fleas. Spittle bugs (also called froghoppers) are in the *Guinness Book of Records* for their leaping ability. It has recently been shown that this enormous acceleration is enabled by resilin-containing tendons in the spittle bug's legs. Like the flea, the leg muscles contract the resilin-loaded tendon and, in a millisecond, all the stored energy is released, propelling the bug into the air at great acceleration.

The resilin pad of the flea under ultraviolet illumination. The resilin pad, which is found in the pleural arch at the top of the hind legs (as labelled by the white box above), is responsible for the flea's jumping powers. Images at the bottom illustrate the pH dependency of the resilin pads fluorescence. The fluorescence is quenched at a lower pH value (i.e. pH 2), but is restored when the pH is brought back up to pH 7 or greater (i.e. pH 12).

According to Weis-Fogh (1961), resilin plays a vital role in saving energy expenditure for insect flight. Insect muscles are able to generate relatively enormous amounts of power (10- to 15-fold more power per unit mass of muscle than a human). However, the power wasted against mass force increases linearly with wing stroke frequency. Therefore, bees, flies and some moths would need to increase the muscular power several times over to facilitate flight and this is beyond the capacity of insect muscles. Insects have solved this problem, through the evolution of a passive energy saving material in their flight systems: resilin.

Flying insects like mosquitoes and bees flap their wings at rates exceeding 300 cycles per second. In its lifetime, a bee will make over 300 million such wing beats. Male cicadas are noted for their loud and high-pitched sound (over 4,000 Hz). The sound is produced in a specialised organ called the tymbal, which is a hollow structure lined with sheets and ribs of flexible cuticle containing resilin. They are the only insects to have developed such an effective and specialised means of producing sound. Some large species produce a noise intensity in excess of 120 dB at close range (this is approaching the pain threshold of the human ear). One small cicada produces sharply resonant sound pulses at over 13·kHz. Dragonflies hover motionless over a single point and can dart at dizzying rates. They use resilin to achieve these aerial feats, both in tendons attached to their wings and in their wing veins. We have also found resilin at the base of sensory hairs in fruit flies, where it likely serves to return the displaced hair to its original position.

Resilin unites these diverse insects and their amazing biomechanical feats. It is an almost perfectly elastic rubber-like protein first identified by Danish researcher Torkel Weis-Fogh, in 1960. The name is derived from the Latin, *resilire*, to jump back. Insects have evolved this exquisitely elastic protein material over hundreds of millions of years, and use it to enhance the efficiency of flight, to permit astounding jumping feats and to produce remarkable songs. Humans and other vertebrates have an elastic protein called elastin (it is found in artery walls and skin). This material, like resilin, must have a very high fatigue lifetime (number of cycles to failure). For example, aortic elastin undergoes millions of stress–strain cycles in a human lifespan. Resilin is the insect world's version of elastin, but it is very different in sequence, structure and function.

42 Dityrosine formation in crosslinked recombinant resilin. a. Structure of the dityrosine adduct. b. Adult dragonfly and extracted wing tendon in PBS viewed under white light (centre) and fluorescing blue under UV light (bottom). c. Moulded recombinant resilin rod viewed under white light (centre) and fluorescing blue under UV light (bottom) d. HPLC analysis of acid-hydrolysed uncrosslinked recombinant resilin and crosslinked recombinant resilin showing the formation of dityrosine following crosslinking.

Naturally cross-linked resilin exhibits two outstanding properties: high rubber efficiency (resilience) and very high fatigue lifetime. Resilience is the property of a material to absorb energy when it is deformed elastically and then, upon unloading, to have this energy recovered.

In its native state, as found in dragonfly tendon, resilin is far superior to other rubbers in its near perfect elastic recovery, having a spring efficiency, or resilience, of >97% (Weis-Fogh, 1960). Superball rubber (polybutadiene) displays a resilience of c. 80%, while inner tube rubber shows only 50% resilience.

In addition, resilin is very 'stretchy', reaching a maximal extension of over 300%, with a low modulus of elasticity, or stiffness. It must survive at least 300 million contraction and relaxation cycles (in cicadas or mosquito wings). Likewise, when a flea jumps, all of the stored energy in its leg muscles is released in around one millisecond. This is roughly the same rate of motion that resilin makes in the wing hinge of a mosquito or in the tymbal organ of a cicada. At the molecular level, protein chains must move very quickly in a near-frictionless manner to achieve this.

In the light of studies revealing the remarkable properties of elastomeric and structural proteins and the rapid advances in molecular biology over the past decade, there has been a surge of interest in reproducing these proteins in a recombinant form. This interest in biomimetics has resulted in a number of biomaterials being produced as recombinant proteins, including spider silks, collagen wheat gliadin, elastin and resilin. In each case, the genes encoding the bioelastomer protein were cloned, or synthesised, inserted into expression vectors (plasmids) and cloned into either eukaryotic (cells with genetic material in a nucleus) or bacterial cells. During their growth phase, the transformed cell lines synthesise large amounts of the recombinant protein, which can be purified from host proteins. The aim is to isolate large amounts of pure recombinant elastomeric protein for subsequent biomechanical,use.

MFKLLGLTLLMAMVVL GRPEPPVNSYLPP
SDS YGAP GQSGPGGRPSDS YGAP GGGNGG
RPSDS YGAP GQGQGQGQGQGG YAGKPSDT
YGAP GGGNGNGGRPSSS YGAP GGGNGGRP
SDT YGAP GGGNGGRPSDT YGAP GGGGNGN
GGRPSSS YGAP GQGQGNGNGGRSSSS YGA
P GGGNGGRPSDT GAP GGGNGGRPSDT YGA
P GGGNNGGRPSSS YGAP GGGNGGRPSDT Y
GAP GGGNGNGSGGRPSSS YGAP GQGQGGF
GGRPSDS YGAP GQNQKPSDS YGAP GSGNG
NGRPSSS YGAP GSGPGGRPSDSYGPPAS G

SGAGGAGGSGPGGADPAKYEFNYQVEDAP
SGLSFGHSEMRDGDFTTGQYNVLLPDGRK
QIVEYEADQQGYRPQIRYEGDANDGSGPS
GPGGPGGQNLGADGYSSGRPGNGNGNGGY
SGGRPGGQDLGPSGYSGGRPGGQDLGAGG
YSNGKPGGQDLGPGGYSGRPGGQDLGRDG
YSGGRPGGQDLGASGYSNGRPGGNGNGGS
DGGRVIIGGRVIGGQDGGDQGYSGGRPGG
QDLGRDGYSSGRPG GRPGGNGQSQDGQG
YSSGRPGQGGRNGFGPGGQNGDNDGGYRY

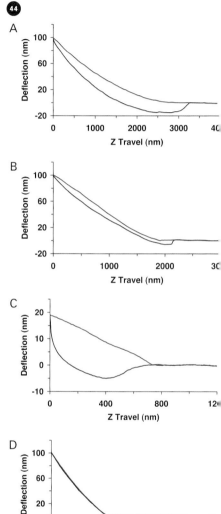

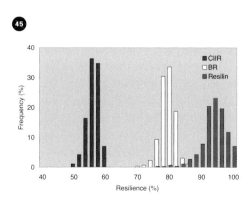

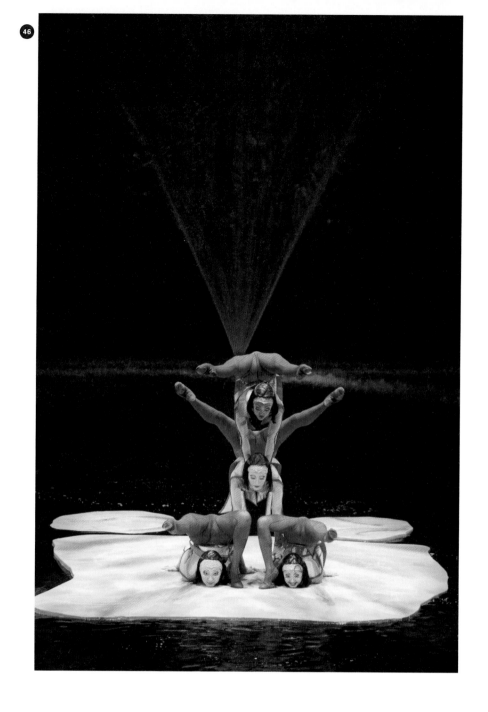

43 The derived amino acid sequence of the *Drosophila melanogaster* resilin gene (CG15920). The repetitive YGAP motifs are highlighted in red. The recombinant resilin protein sequence is highlighted in grey. This represents the first exon of the CG15920 gene and contains the elastic domain of the protein. The signal sequence is shown in purple and was not included in the recombinant clone construct. The single-letter amino acid codes (IUPAC-IUB nomenclature) are shown.

In 2001, Ardell and Anderson published a putative sequence of the fruit fly (*Drosophila melanogaster*) resilin gene. Using this knowledge, we were able to clone and express the recombinant resilin protein (see figure 43) in bacteria and we introduced dityrosine crosslinks into the soluble form of the recombinant protein via a facile photochemical method. We also succeeded in scaling-up the production of recombinant resilin using an optimised fermentation procedure and to simplify the purification of recombinant resilin protein. We have also cloned a purely synthetic version of the mosquito resilin gene, AN16 and have shown that the recombinant protein expressed from that construct also displays high resilience and can be crosslinked via dityrosine covalent bonds using the photochemical method applied in the synthesis of recombinant resilin. With access to this man-made material we can begin to dissect the structural basis for resilin's remarkable mechanical properties. Recent studies have demonstrated the unstructured and highly flexible nature of the resilin protein chains, consistent with a random-network elastomer model, for elastin, which requires a high degree of disorder within the polymer chains.

Tensile tests as well as nano-mechanical measurements using a scanning probe microscope showed the recombinant resilin to be just as resilient as the natural material and more resilient than butadiene (superball) rubber (Figures 44 and 45). Both natural and recombinant resilins contain large quantities of water, which plays an important role in imparting the unique properties of superior elasticity on the materials. When the recombinant resilin was allowed to dry out it lost its resilience (Figure 44) and became stiff and leathery. The modulus or stiffness of the recombinant resilin is much lower than the natural material. This has been determined to be due to a much lower molecular mass in the recombinant material and the likely unproductive nature of some of the intramolecular dityrosine bonds formed during photochemical crosslinking.

44 Elastic properties (resilience) of elastomers measured with a scanning probe microscope illustrating the difference between a low resilience rubber (chlorobutyl, CIIR) and a highly resilient rubber (butadiene, BR) as well as the importance of water and complete protein hydration for the high resilience of resilin. Typical force curves (Deflection vs Z travel) are shown for (a) chlorobutyl rubber, (b) butadiene (superball) rubber, (c) partially dried crosslinked recombinant resilin at 80% relative humidity and (d) fully hydrated recombinant resilin. The approach curve is shown in red and the retract curve in blue.

Although nature is inherently better at materials development and design than scientists and engineers, we can attempt to understand and ultimately manipulate natural materials by investigations based on the interface between biology, polymer chemistry and engineering. We are continuing to study the biochemical and physical properties of the recombinant and synthetic forms of resilin and aim to synthesize chemical mimics of this remarkable protein. The knowledge generated from research into resilin structure and function will underpin the improved design of novel biomaterials leading to suggestions for improved specialty polymers and materials for a wide range of possible applications, including biomedicine and industry.

Resilin has a remarkably high fatigue lifetime and our aim is to reproduce this desirable mechanical property in synthetic materials. We believe that resilin-like materials may be used in the future in the medical device field as components of prosthetic implants, including spinal disks. Materials designed for this purpose must survive for at least 100 million cycles of contraction and relaxation in a lifetime.

45 Scanning probe microscopy results comparing the resilience of crosslinked recombinant resilin (fully hydrated) to chlorobutyl rubber (CIIR) and butadiene rubber (BR).

References

Aaron, B.B. and Gosline, J.M. (1981) 'Elastin as a random-network elastomer: A mechanical and optical analysis of single elastin fibers'. *Biopolymers* 20:1247–1260.

Andersen, S.O. (1964) 'The crosslinks in resilin identified as dityrosine and trityrosine', *Biochim. Biophys. Acta* 93: 213–215.

Andersen, S.O. (1966) 'Covalent cross-links in a structural protein, resilin', *Acta Physiol. Scand. Suppl.* 263: 1–81.

Andersen, S.O. (2004) 'Regional differences in degree of resilin cross-linking in the desert locust, *Schistocerca gregaria. Insect Biochem. Mol. Biol.* 34: 459–466.

Ardell, D.H. and Andersen, S.O. (2001) 'Tentative identification of a resilin gene', in *Drosophila melanogaster. Insect Biochem. Mol. Biol.* 31: 965–970.

Bennet-Clark, H.C. and Lucey, E.C.A. (1967) 'The jump of the flea: a study of the energetics and a model of the mechanism'. *J. Exp. Biol.* 47: 59–76.

Bennet-Clark H.C. (2007) 'The first description of resilin'. *J. Exp. Biol.* 210: 3879–3881.

Burrows, M. (2003) Biomechanics: froghopper insects leap to new heights. *Nature* 424: 509.

Burrows, M., Shaw, S.R. and Sutton, G.P. (2008) 'Resilin and cuticle form a composite structure for energy storage in jumping by froghopper insects'. *BMC Biology* 6: 41.

Elmorjani, K., Thievin, M., Michon, T., Popineau, Y., Hallet, J.N., and Gueguen, J. (1997) 'Synthetic genes specifying periodic polymers modelled on the repetitive domain of wheat gliadins: Conception

CAN'T BELIEVE I'M FLEXIBLE LIKE THIS

The contortion team at 'O' is made up of six girls. They have two scheduled rehearsals each week. According to Odmaa, performing two shows a night, five nights a week, they do not need to rehearse more than that.

Odmaa practices pilates which is good for the back, joints and different muscle groups. Running and weight lifting is not good for the muscles used by the contortionists.

She has no special dietary requirements and tends to eat 2–3 hours before performing. She says "I feel lighter so it is easier to perform".

When Odmaa was young, she used to sit and sleep in unusual positions. She saw contortionists performing on TV and thought "I could do that!" Her parents always knew that she was different.

She grew up in small town and had to travel to the capital city of Mongolia to meet with a contortion coach. After meeting with the coach, she went back home with 'homework' to practice what she learned.

With regard to her elasticity, Odmaa says, "I have no idea why. I can't believe that I'm flexible like this!"

and expression'. *Biochem. Biophys. Res. Commun.* **239**: 240–246.

Elvin, C.M., Carr, A.G., Huson, M.G., Maxwell, J.M., Pearson,R.D., Vuocolo,T., Liyou, N.E., Wong, D.C.C., Merritt, D.J. and Dixon, N.E. (2005) 'Synthesis and properties of crosslinked recombinant pro-resilin'. *Nature* **437**: 999–1002.

Flory, P.J. (1953) *Principles of polymer chemistry.* Cornell University Press, Ithaca, New York.

Flory, P.J. and Rehner, J. Jr. (1943) 'Statistical mechanics of cross-linked polymer networks. I. Rubberlike elasticity'. *J. Chem. Phys.* **11**: 512–520.

Fonseca, P.J. and Bennet-Clark, H.C. (1998) 'Asymmetry of tymbal action and structure in a cicada: a possible role in the production of complex songs'. *J. Exp. Biol.* **201**: 717–730.

Gorb, S.N. (1999) 'Serial elastic elements in the damselfly wing: mobile vein joints contain resilin'. *Naturwissenschaften* **86**: 552–555.

Gosline, J., Lillie, M., Carrington, E., Guerette, P., Ortlepp, C. and Savage, K. (2002) 'Elastic proteins: biological roles and mechanical properties'. *Philos. Trans. R. Soc. Lond B Biol. Sci.* **357**:121–32.

Gough, J. (1805) 'A description of a property of Caoutchouc or India rubber; with some reflections on the cause of the elasticity of this substance'. *Memoirs Lit. and Phil. Soc. Manchester* **1**: 288–295.

Huson, M.G, and Maxwell, J.M. (2006) 'The measurement of resilience with a scanning probe microscope'. *Polym. Testing* **25**: 2–11.

Huson, M.G. and Elvin, C.M. (2008) 'Recombinant Resilin—a protein based elastomer': *Current Topics of*

Elastomer Research. Taylor & Francis Group LLC, CRC Press, Florida, USA.

Ito, H., Steplewski, A., Alabyeva, T. and Fertala, A. (2006) 'Testing the utility of rationally engineered recombinant collagen-like proteins for applications in tissue engineering'. *J. Biomed. Mater. Res. A.* **76**: 551–560.

James, H.M., and Guth, E. (1943) 'Theory of the elastic properties of rubber'. *J. Chem. Phys.* **11**: 455–481.

Kim, M., Elvin, C., Brownlee, A. and Lyons, R. (2007) 'High yield expression of recombinant pro-resilin: lactose-induced fermentation in *E. coli* and facile purification'. *Protein Expr. Purif.* **52**: 230–236.

Lazaris, A., Arcidiacono, S., Huang, Y., Zhou, J.F., Duguay, F., Chretien, N., Welsh, E.A., Soares, J.W. and Karatzas, C.N. (2002) 'Spider silk fibers spun from soluble recombinant silk produced in mammalian cells'. *Science* **295**: 472–476.

Lyons, R.E., Lesieur, E., Kim, M., Wong, D.C., Huson, M.G., Nairn, K.M., Brownlee, A.G., Pearson, R.D., Elvin, C.M. (2007) 'Design and facile production of recombinant resilin-like polypeptides: gene construction and a rapid protein purification method'. *Protein Eng. Des. Sel.* **20**: 25–32.

Miao, M., Cirulis, J.T., Lee, S. and Keeley, F.W. (2005) 'Structural determinants of cross-linking and hydrophobic domains for self-assembly of elastin-like polypeptides', *Biochemistry* **44**: 14367–14375.

Nairn, K.M., Lyons, R.E., Mulder, R.J., Mudie, S.T., Cookson, D.J., Lesieur, E., Kim, M., Lau, D., Scholes, F.H. and C.M. Elvin (2008) 'A synthetic pro-resilin protein is largely unstructured in solution'.

Biophys. J. **95**: 3358–3365.

Rothschild, M. and Schlein, J. (1975) 'The jumping mechanism of *Xenopsylla cheopis*. I. Exoskeletal structures and musculature'. *Philos. Trans. R. Soc. Lond. B Biol. Sci.* **271**: 457–490.

Rothschild, M., Schlein, J., Parker, K., Neville, C. and Sternberg, S. (1975) 'The jumping mechanism of *Xenopsylla cheopis*. III. Execution of the jump and activity'. *Philos. Trans. R Soc. Lond. B Biol. Sci.* **271**: 499–515.

Rothschild, M. and Neville, C. (1967) 'Fleas—insects which fly with their legs'. *Proc. R. Entom. Soc. Lond. C.* **32**: 9–10.

Staudinger, H. (1920) 'Über polymerisation'. *Berichte der Deutschen Chemischen Gesellschaft* (A and B Series) **53**: 1073–1085.

Taksali, S., Grauer, J.N. and Vaccaro, A.R. (2004) 'Material considerations for intervertebral disc replacement implants'. *Spine J.* **4**: 231S–238S.

Tatham, A.S. and Shewry, P.R. (2002) 'Comparative structures and properties of elastic proteins'. *Phil. Trans. R. Soc. Lond. B* **357**: 229–234.

Treloar, L.R.G. (1975) *The physics of rubber elasticity.* Clarendon Press, Oxford.

Wall, F.T. (1942) 'Statistical thermodynamics of rubber'. II. *J. Chem. Phys.* **10**: 485–488.

Weis-Fogh, T. (1960) 'A rubber-like protein in insect cuticle'. *J. Exp. Biol.* **37**: 889–907.

Weis-Fogh, T. (1961) Power in flapping flight. In:*The Cell and the Organism.* Ramsay, J.A. and Wigglesworth, V.B. (eds). Cambridge University Press, Cambridge.

THE LONELIEST MONK

The music is performed with pipa and ten western instruments.
It was inspired by a Chinese monk and American jazz legend
Thelonious Monk.
Premiered on May 21, 2004 at the Kitchen in New York with the
Kitchen House Blend.

RESONANCE

Bending without breaking, elastic microtonality—in Indian music, one bends notes to find their essences. Stretching a material to its extreme reveals something essential about its true nature. But we need a container within which to experience music, an architectural instrument, a cage perhaps, one which simultaneously constrains and through its restriction allows the trajectories of resonance to take form. Music doesn't exist without a medium within which sound can vibrate elastically and the quality of the resonance in turn helps us decide about the quality of the architecture. Couldn't this be a model for an 'elastic city'? An entire city could be built with these sound architectures vibrating in an urban environment where navigational trajectories are so designed as to harness the full creative energy of its inhabitants.

Resonate

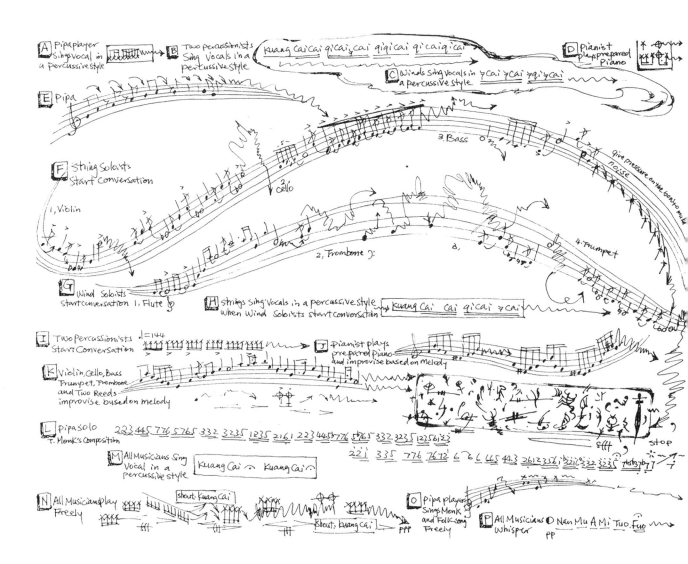

DITTY FOR ONE STRING

You have touched this string
and it rang

it rang
and now it seems to be silent

yet the sound goes on
inaudible
strangely embodied in
all that there is

this durable tune
the singing silence

The god of universal elasticity
now looks at his fingers and counts
he takes his abacus, opens his computer
unravels slender chains of thought,
tendrils of figures and signs,
winds them around this
happening of the heart,
this little playful happening
which he tries to catch, to pin down
to put in his book:

this string
its motionlessness
and musical silence.

He checks again stress and strain
yes:
the string shall be restored
like on day one,
it is
of course
reversible

but he knows that cannot be
after you have struck
a string
and made it ring or made it cry
or muted it in-between

he knows too well that cannot be
—not wholly
not as predicted.

So he returns to safe ground
and says: first you strike the string
and it shall be heard
and then it shall relapse into silence
unchanged, unharmed and in tune
as if nothing has happened

he says so and remembers reasons
to be proud—he, the protector of his flock
through hard times
from the beginning:
bend
and thou shalt not break
bend
thus thou shalt not break.

His book says more of the story
of things and peoples,
our own peoples—
about yielding
and limits
about shattered stars
and paradoxes of healing.

Pluck
a string
so it may ring

ring
it will
and never relapse
into the same silence

IN SILENCE

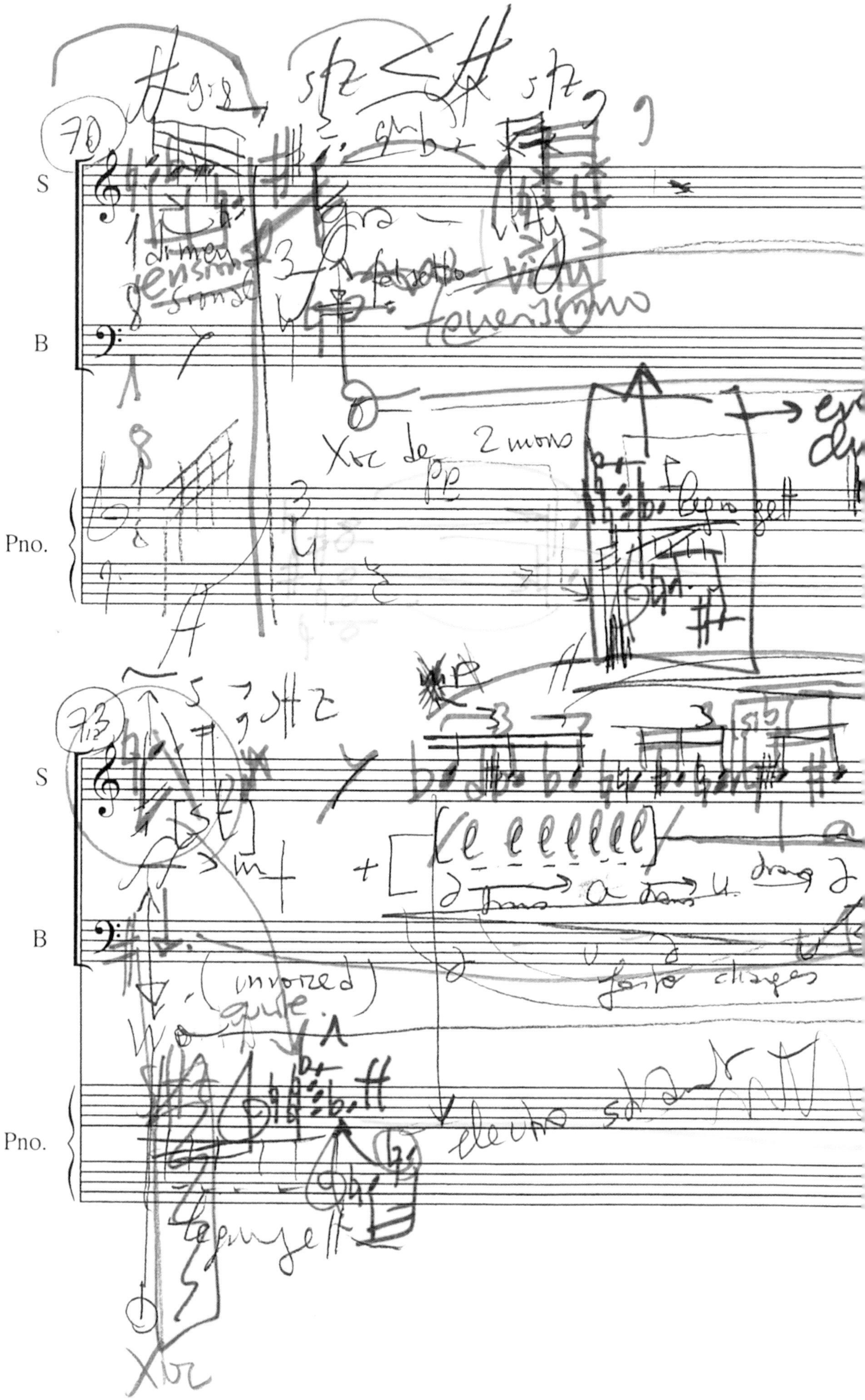
S
B
Pno.
S
B
Pno.
2 mono
falsetto
(unvoiced)

HYPERMUSIC PROLOGUE

First draft of a page of the 4th Plane for *Hypermusic Prologue*, with libretto by physicist Lisa Randall and stage design by Matthew Ritchie. In this first musical pencil-draft, I composed the soprano and baritone melodies, and orchestral harmonies with different colours representing different musical 'materials': the deep blue for the vivid and sharp soprano expression, the orange for the long, very tender but desperate baritone, sky blue for some soft string harmonies and red-magenta for sharp and tense wind-percussion interventions.

Hypermusic Prologue is a collaboration between science, music and art. It explores the historic form of opera to generate a dramatic expression of 21st century ideas, including recent developments in higher dimensional physics and their parallels in music and art.

The soprano, experiences a deep tension between her human love and her passion for knowledge. This tension is realized dramatically through the limited spacetime experience she shares with her lover (baritone) and her belief that there is a larger world to be explored. Their interaction is disturbed and illuminated when the soprano embarks on a hypothetical trip into a warped extra dimension. Their alternative views and experiences of reality take on metaphorical meaning through this journey.

A SPIDER'S WEB

If you take a walk on a foggy morning in a meadow or head out into your own garden you will notice them everywhere, and see how beautiful they are. The functional elegance and structural complexity of the two-dimensional orb web is highlighted by the thousands of tiny dew droplets attaching to the silk threads like silvery beads on a string.

If you stop for a while and look closer at the web, you will see that it consists of more or less distinct parts. In the centre, or to be accurate somewhat north of the geometric centre as most orb webs are asymmetric, there is the hub, where, depending on the species, the spider may or may not be found. If it is not in the hub you will most likely see a signal thread leading from the hub to nearby vegetation where the spider will be holed up in its hiding place. From the hub you will see the radii running out like spokes on a wheel. They function both as support for the capture spiral and as an information highway for the vibrations emitted from the struggling prey. They provide the spider with information on the location and size of the unfortunate insect.

Due to its pivotal role in prey capture, the capture spiral takes up most of the space, consumes most of the silk used and most of the time needed to build the web. A frame from which anchor threads attach the web to its surroundings, encloses the radii and the capture spiral. If you follow these anchor threads you will notice that some of them can be very long, allowing the web to be placed out in the open.

Not only are webs beautiful to look at. They are also very impressive from an engineering perspective. A web weighing less than 1 mg is capable of stopping a 25 mg fly with an incoming speed of up to 4 m/s, or under the right circumstances even much larger insects. If you could increase the frame rate of your visual system and happened to be standing aligned to the web plane at the exact moment where the fly flies into it, you would see how the impact deforms the web out of the plane before the individual strands pull it back again. This behaviour is mainly caused by the amazing material properties of spider silk. It has a low weight and a high elasticity, but at the same time has a remarkably high tensile strength, which combined give spider silk a strength-to-weight ratio five times greater than steel and a toughness three times greater than Kevlar.

WHY ELASTICITY IS BAD

The spider has one problem with a very elastic web. The stretching of the web as a prey hits it stores elastic energy and this energy needs to be released. Imagine the web like a trampoline. If you throw something at a trampoline it will bounce right back. A web that reacts to a fast incoming fly by shooting it out in the opposite direction is no good for the hungry spider. The problem is solved by making the silk imperfectly elastic, so that more energy is dissipated by internal friction and heat production in the individual silk threads and by aerodynamic damping of the whole oscillating capture spiral, while at the same time ensuring that the prey sticks to the silk by (in the majority of spiders) coating the threads with glue. A separate, smaller group of orb spiders uses a tangle of very fine threads which adhere to the prey by Van der Waals and hygroscopic forces.

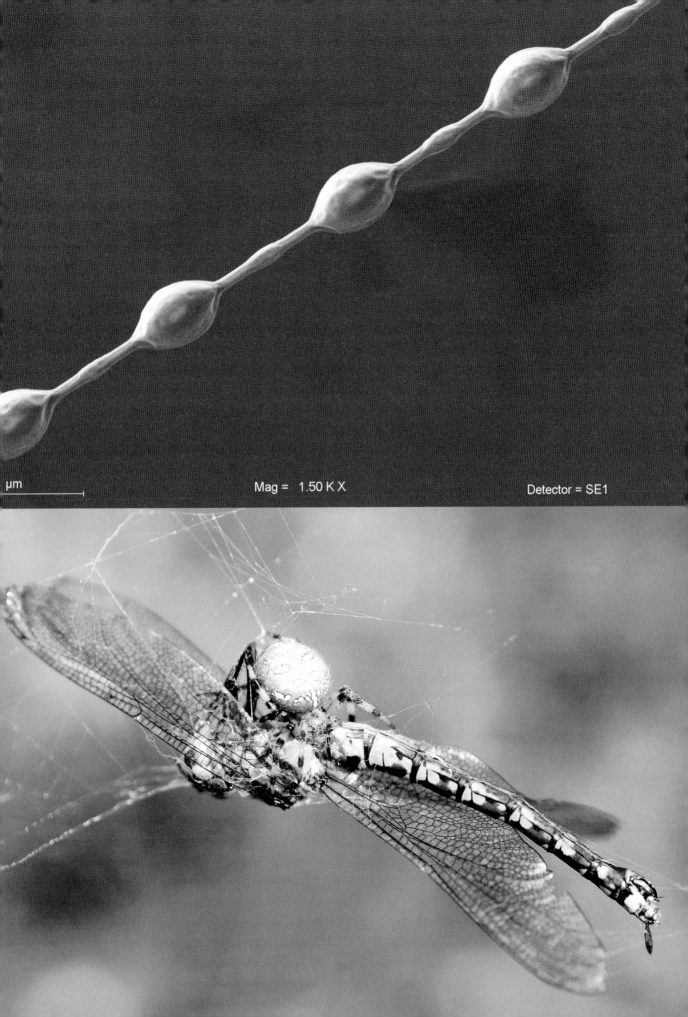

Mag = 1.50 K X

Detector = SE1

WHY ELASTICITY IS GOOD

The kinetic energy from the incoming insect has to be absorbed somehow. This is best done in an elastic structure. In a rigid structure the incoming energy would have to be absorbed by the individual threads making them much more likely to break. To avoid this, much thicker threads and much more glue would be required making its construction too energetically expensive. A further advantage of elastic threads is that they make it more difficult for the prey to break free. Just compare how much more easily you walk on solid ground, which gives good resistance to your downward force, than for instance on an inflatable mattress, which deforms when you step on it. The trapped prey has the same problems with generating enough force to break free of the web. Finally, when the web can deform out of plane, it faces a reduced risk of being destroyed by strong winds.

WHY WE DO NOT LIKE ELASTICITY

During the last 30 years the effort to construct artificial spider silk has intensified. A low weight material with high strength and toughness would be very useful. One of the envisaged uses is to replace Kevlar in bulletproof vests. However, elasticity is not much of an advantage here. On the contrary, if you are wearing a bulletproof vest made of spider silk and somebody shoots you then there would have to be a space between your body and the vest of perhaps 10-20 cm allowing the vest to deform before stopping the bullet. So the challenge is to develop artificial spider silk with less elasticity than the real thing. One potential biomimetic application of the spider web would be as a lightweight structure deployed in space and assembled from easy-to-transport parts, perhaps by a small spider-like robot. In this way a large area can be covered using only a small amount of material with a small mass—something of crucial importance considering the very high price per kg of sending things into orbit. However, there is an obvious drawback to this: a construction based on spider web elasticity will collapse in empty space, whereas a construction of rigid rods or wires would not.

WHY WE MIGHT LIKE ELASTICITY AFTER ALL

A recent study has shown that spider silk not only shows length changes because of prey or wind impact, but also as a result of changes in humidity.[1] Furthermore, the study demonstrated that the force generated by the cyclic contraction and relaxation of the individual fibres in response to varying humidity had a very high power density, thus making it potentially suitable as a biomimetic muscle. These have a large potential application value in biomedical engineering and biomechanics, where, due to their biological compatibility and low working temperatures, they can be used in artificial limbs and organs. They would also be of great value in industry in general and in advances robotics in particular. It is very likely that more advantages of using elastic and compliant materials such as silk will become apparent in the coming years. One thing is almost certain: the orb web and its constituent silk have many more secrets yet to reveal.

1 J. Exp. 2009, vol. 212, pp. 1990-1994

CHICKEN NUGGETS AND BABIES

I stretched my horizons with alcohol when I felt trapped. When I was a child I often could not breathe, knowing that there was more out there, while at the same time, being told that I needed to be sensible, settle down, get a job at the postal service. I learned to travel inside my body, with drugs... food... no food... love... thoughts...

I have seen other girls with limited opportunities stretch their fortunes with babies... sex... puffed up rage... love... Wal-Mart... Chicken Nuggets... babies.

Stretch

DATE	NUMBER	CHARGE	DISPOSITION
12/20/75	#8948	240.20-3 PL Disorderly Conduct	2/9/76 - Fined $50.
2/9/76	#9118	Bench Warrant Failed To Appear	See Above
6/4/76	#9529	240/25-5 PL Harassment	12/31/76 - Dismissed
10/25/76	#10,028	Family Court Warrant	see above T O T Family Court
8/5/77	#11,485	155.25 PL Petit Larceny	9/14/77 red. 240.20 PL Dis. Con./ fined $10.00
		221.05 PL Poss. Marijuana	9/14/77 Cond. Discharge

(snap shot of me at 12)
I made these pants because I
love Janis Joplin. She saved
my life. The irony.

black and white photo of me in my 20s
when I worked as a snake charmer in
the carnival sideshow

Arrest record

The map of the world in my first grade classroom at Saint James Institute. When I was little, I used to bang my head on the wall hoping that I would make a hole big enough for me to crawl through.

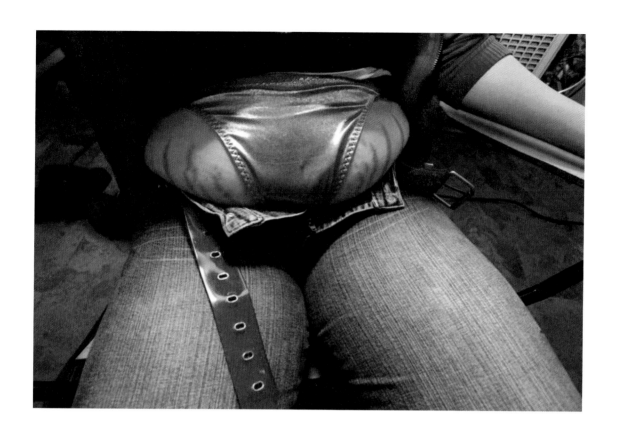

Chocolate cream pie

Dana on her Wedding Day

One month after baby

Heather, 24 and 5 months pregnant

53

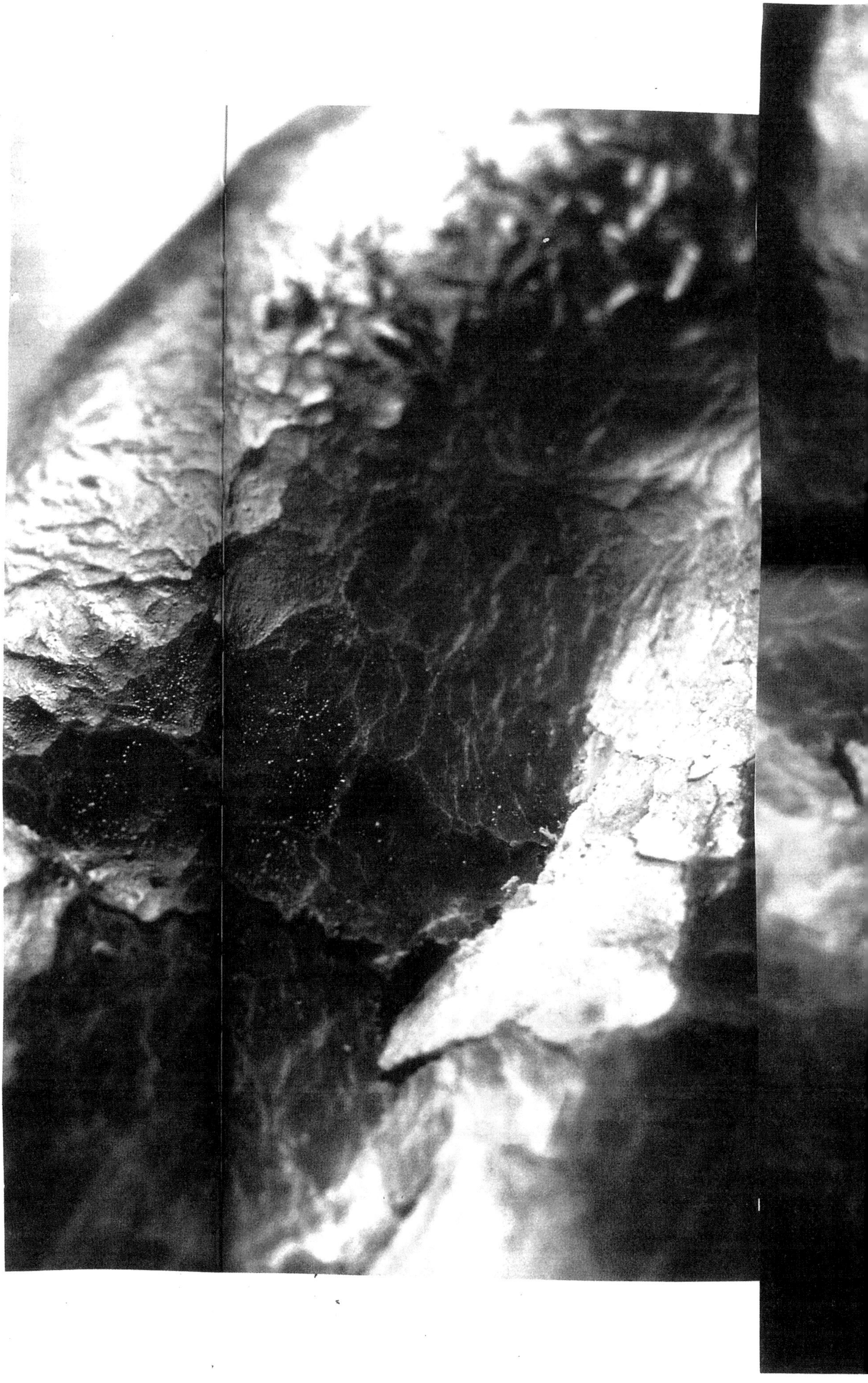

76 <u>INT/GALLERY/DAY</u>

OMITTED

77 <u>INT/~~AMBULANCE~~/EVENING</u>

Ditte is sitting in an ambulance with Dad whose eyes are closed, Ditte's got her arm around him

The ambulance slows down, turns into the yard behind the bakery

> DITTE
> Dad?

> DAD
> Yeah

> DITTE
> We're here now

Dad looks out the window, nods

78 INT/BAKERY/EVENING

... bakery, Ditte is right behind them, they ... him, Eigil comes over and shakes

... looking at what's left of their boss. ... Hi, Rikard

... fingers, go ahead and bake, the bakers ... the air, bread is taken out, dough is

... roud, loves his bakery, loves bread, the smell, the flour

79 <u>INT/DAD'S LIVING ROOM/EVENING</u>

~~Orderly helps Dad back into bed, Ditte turns on a single lamp, Ditte nods a thank you to the Orderly who leaves, Ditte tucks Dad in, he doesn't react~~

Dad's eyes are swimming, he sees Ditte, his breath is catchy, whispers something inaudible, Ditte puts her ear close to his mouth, but nothing comes out, then she takes over:

> DITTE
> You're my dad, and I'm your daughter

Dad looks at her

> DAD
> Chrisser ... You have to take care of her

> DITTE
> (I love Chrisser)

Dad nods

> DITTE
> I love you

Dad nods

> DITTE
> You love me too

Ditte struggles to say the right thing, but nothing else comes out, Dad just whispers:

> DAD
> Ditte Rheinwald

> DITTE
> ~~Ditte Rheinwald, that's me~~

She nods. She waits, but there's no more, he seems to have dozed off now, she gets up and whispers:

> DITTE
> I'm going home now, Dad

He focuses on her, they both know soon they'll be parting for good, he manages to whisper:

> DAD
> Yes ... home to Peter

> DITTE
> That's right. I'm going home to Peter

She strokes his cheek

> DITTE
> ~~Do you want me to turn out the light?~~

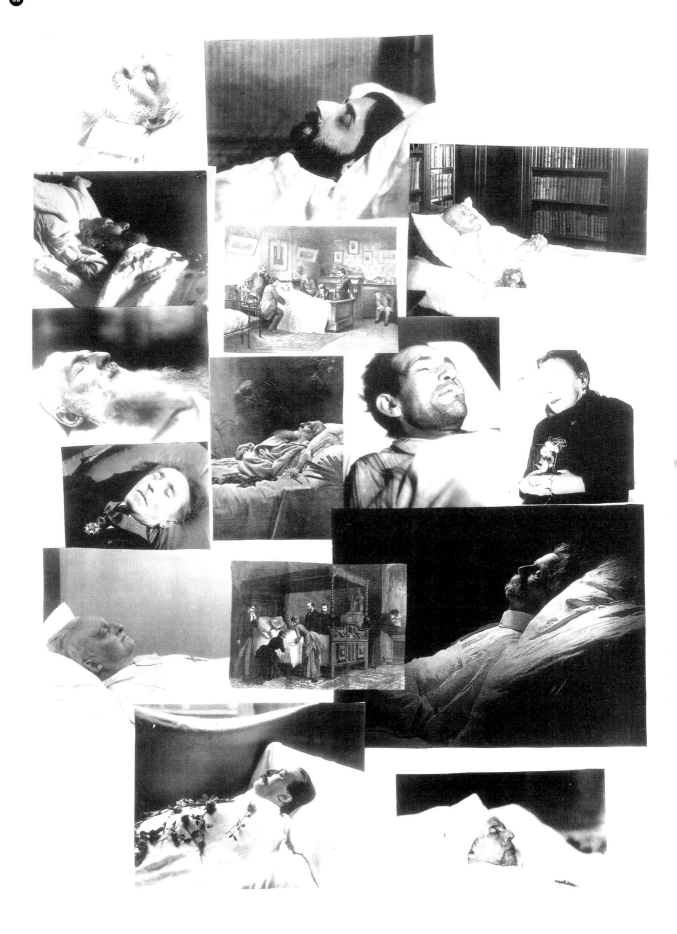

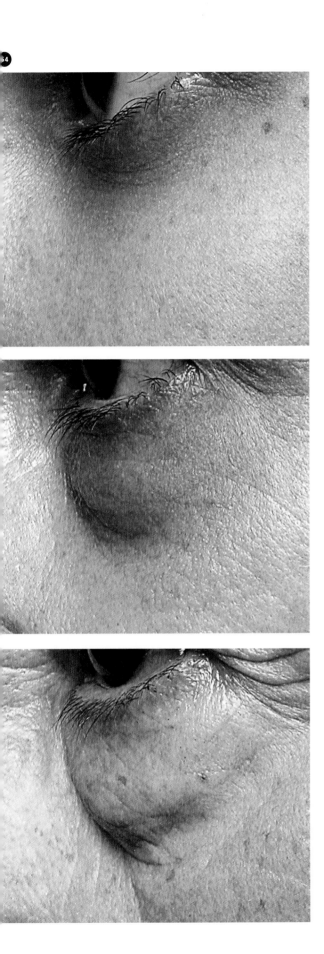

Rodin sur son lit de mort, 1917
Paris, musée Rodin
Photographer: Harry B. Lachmann

Marcel Proust sur son lit de mort, 1922
Paris, musée d'Orsay
Photographer: Emmanuel Radnitsky, dit
Man Ray, 1890–1976

Léon Blum sur son lit de mort, 1950
Paris, Bibliothèque nationale de France,
département des Estampes et de la Photographie
Anonymous photographer

Léon Gambetta sur son lit de mort, 1882
Paris, Bibliothèque nationale de France,
département des Estampes et de la Photographie
Photographer: Étienne Carjat, 1828-1906

Mort de Gambetta aux Jardies dans la nuit
du 31 décembre 1882 au premier janvier
1883
Paris, Bibliothèque nationale de France,
département des Estampes et de la
Photographie
Photographer: H. Marrès

Édith Piaf sur son lit de mort, 12 octobre 196.
Paris, musée-association Les Amis d'Édith
Piaf
Anonymous photographer

Rodin sur son lit de mort, 1917
Paris, musée Rodin
Photographer: Harry B. Lachmann

Le Grand Sculpteur Rodin sur son lit de
mort, 1917
Paris, collection particulière
Photographer: J. Clair-Guyot

Chaïm Soutine sur son lit de mort, 1943
Paris, Bibliothèque nationale de France,
département des Estampes et de la
Photographie
Photographer: Rosa Klein, dite Rogi André,
1905-1970

Jean Cocteau sur son lit de mort, 1963
Paris, Mission du patrimoine
photographique
Photographer: Raymond Vionquel,
1912–1994

Eugène Carrière sur son lit de mort, March 1906
Paris, musée d'Orsay
Photographer: Félix Tournachon, dit Nadar,
1820–1910

Georges Hugo sur son lit de mort, 1925
Paris, maison de Victor Hugo
Photographer: H. Gerschel

Les Derniers Moments de Victor Hugo, 30th of May 1885
Paris, maison de Victor Hugo
Photographer: E. H. Tilly

Gustave Doré sur son lit de mort, 1883
Paris, Bibliothèque nationale de France,
département des Estampes et de la Photographie
Photographer: Félix Tournachon, dit Nadar,
1820–1910

Jean-Auguste Dominique Ingres sur son lit de mort, 1867
Montauban, musée Ingres
Photographer: Charles Marville, 1816–1878

SOPHIE SEITÉ, CATHERINE GERST, BIOCHEMISTS

SKIN (ELASTICITY) CHANGES AS SKIN AGES

AGE AND THE APPEARANCE OF AGE

Cutaneous ageing, from a cosmetic and dermatological point of view, is particularly interesting because of some unique characteristics of the skin: it covers the whole body and is the interface between the external environment and the organism. It is also subject to two types of ageing: intrinsic or chronological ageing due to age and genetic factors and extrinsic ageing due to the environment and its main factor, the sun.

This means that age and the appearance of age are separate entities and that significant disparity may be observed between the two. Chronological ageing affects all parts of the body but it is best assessed on covered areas not exposed to the sun. Apparent age results from the superimposition of chronological ageing and photo-ageing and can easily be appreciated in areas exposed to the sun such as face, neck or arms.

The passing of time and the environment not only affect the appearance of the skin but also its function. The skin is not a simple envelope: it is subjected to mechanical strains which it withstands due to its flexibility. With time, its properties of elasticity and tension are lost; its protective, metabolic and sensory functions (as well as its capacity to adapt) diminish.

CLINICAL SIGNS OF SKIN AGEING

Modifications of the skin protected from the sun are relatively moderate: thinning of the skin, fine lines, skin dryness, laxness or loss of firmness, benign epithelial proliferation and senile angiomas.

Wrinkles or expression lines are the permanent external signs of the concertina-like action of the skin and the normal furrows of the face constantly produced by the action of the platysma muscles. The sagging folds follow from the loss of tone of the skin's muscles related to the combined action of dermo-hypodermal alterations and gravity, responsible for the pull downwards of the subcutaneous tissue and in particular of fat tissue. The consequence is the appearance of heavy cheeks, a double chin, pouches under the eyes and ptosis of the eyelids.

Pouches under the eyes appear with age.

The pinch test, or the skin fold test, is easily to perform and may help you to evaluate your skin elasticity.
Gently pinch the back of your hand, release, and watch closely to see how long the skin takes to regain its normal aspect (no more fold).

It can be noted that the fold of the skin persists even for as long as 20 seconds in the elderly, indicating a loss of the elastic quality of the skin.

Modifications in skin exposed to the sun add to the changes previously described and are characterised by two essential features: actinic elastosis and the appearance of marks. Elastosis appears as a uniformly coarse skin which is rough and yellowish (citrine skin). The skin is loose, covered with wrinkles and furrows. The rhomboid appearance (from the Greek rhombos, lozenge) of the skin of the nape of the neck of people who have worked all their lives outside (sailors, farmers, builders etc.) is a striking illustration of this. The skin is marked by deep furrows which intersect and outline lozenges of variable size.

56 'Cutis Rhomboidalis' nuchae: the deep wrinkles observed in skin exposed repeatedly to the sun are related to a disorganisation of the network of elastic fibres in the dermis.

EVALUATION OF THE MECHANICAL PROPERTIES OF THE SKIN

57 The twistometer measures skin elasticity. It twists the skin in a zone bounded by the ring and the resulting deformation of the skin (dependent on the elasticity) is measured relative to time.

The results obtained by this type of measurement show that in the elderly the skin is more rigid, more difficult to deform or stretch and that elastic recovery after deformation is reduced on any skin site. This reduction can be up to 30% on the forearm. The decrease in skin mechanical properties is a function of increasing age and external abuses (i.e. sun exposure).

STRUCTURE OF THE SKIN

Anatomically, the skin is an organ composed of three compartments: the epidermis, the dermis and the hypodermis. In an adult, the skin weighs between 3.5 and 4.5 kilograms and its total surface area can be up to two square metres. Its thickness varies between 1.5 and 4 mm depending on the region of the body, being thick on the soles of the feet or on the palms of the hands and thin on the eyelids.

This envelope, combining properties of suppleness and strength, not only protects the organism from external attack but also represents a site for exchange between the inside of the body and the exterior.

58 cross-section of skin

THE DERMIS: A NETWORK STRUCTURE

The dermis makes up a large part of the thickness of the skin. It is the supporting tissue of the skin, compressible, extensible and elastic. Its properties are provided by its architecture, which results from interactions between the constituents of the extracellular matrix and the fibroblasts, both synthesising them and breaking them down. Thus, collagen and elastin fibres form a fibrous web which is bathed in a gel of macromolecules (proteoglycans and glycosaminoglycans), while structural glycoproteins provide the interface between the fibroblasts and the extracellular matrix which surrounds them.

The dermis is subdivided into two areas: a narrow superficial area or papillary dermis and a deep area or reticular dermis that makes up more than 80% of the total thickness of the dermis.

The papillary dermis in a young skin is characterised by the presence of dermal papillae with the appearance of a mountain range. It is a loose connective tissue formed from collagen fibres organised in bundles and

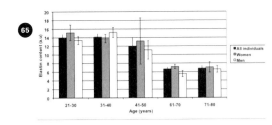

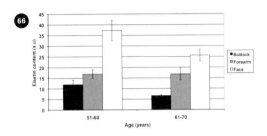

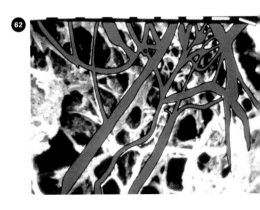

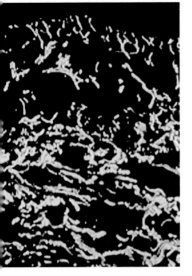

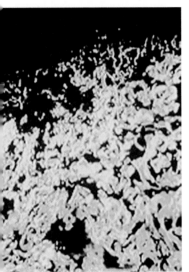

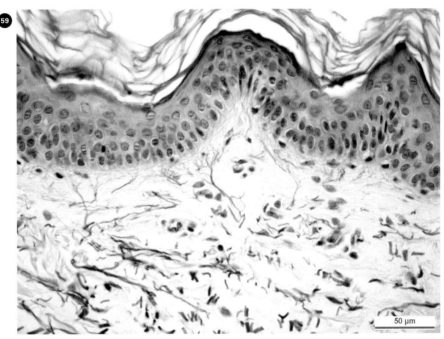

elastic fibres composed of oxytalan and elaunin fibres. The oxytalan fibres, anchored to the dermo-epidermal junction and vertically orientated relative to the surface of the skin, give the papillary dermis its elasticity. It is in this area irrigated by blood and lymphatic capillaries that nutrient exchange takes place with the non-vascularised epidermis.

The reticular dermis is composed of collagen fibres arranged in waves which criss-cross each other in all directions but always remain parallel to the skin surface. The collagen associates with elastin fibres in a thicker fibre network which is more and more compact the deeper you penetrate. It also contains arterioles, venules, small nerves, pilo-sebaceous follicles and the excretory channels of the sweat glands.

59 Histological section of young skin not exposed

and

60 an aged skin exposed to sun.

Modifications of the elastic fibre network are responsible for the lax, sagging appearance of the skin and for the fine wrinkles characteristic of chronological ageing. In the papillary dermis, the vertically orientated candelabra-like network of fine oxytalan fibres, mounting towards the dermo-epidermal junction gradually disappears. This causes flattening of the dermal papillae with the consequence of a loss of epidermal tone, indicating the beginnings of wrinkles. So this level in the skin is one where cosmetic products could be beneficial, and it is why when we evaluated the efficacy of an anti-ageing product we studied its effect on the dermo-epidermal junction.

61 Sun-exposed skin

62 Non-exposed skin

The elastic fibres (shown in blue) are vertically oriented in non-exposed skin and horizontally oriented in sun-exposed skin.

SOLAR ELASTOSIS: THE HISTOLOGICAL SIGN OF PHOTO-AGEING

Analysis of photo-aged skin shows major modifications in the dermis: deterioration of the connective tissue, decrease in the collagen content, the bundles of which appear less numerous, more spindly and disorientated; and accumulation of degenerated elastic fibres known as 'solar elastosis'. The reduced expression of type VII collagen, a major component of dermo-epidermal junction anchoring fibrils, contributes to the formation of wrinkles and to the fragility of the photo-aged skin.

The elastin fibres synthesised by the fibroblasts in the dermis are broken and altered by the repeated action of sun exposure. These abnormal fibres can no longer be digested by the elastase enzymes, because a protein, lysozyme, has been deposited on these fibres. They clump together to form little white balls visible under the skin. In addition, the overall collagen content is decreased and the ratio between type I collagen (the biggest proportion in young skin) and type II collagen is modified, because of an over-stimulation of type I collagenase which specifically degrades type I collagen. The skin, which has lost its elasticity, becomes loose and deep furrows mark the flesh: 'solar elastosis'. It is very difficult to repair this long-term impact; prevention by using daily photoprotection is definitely preferable.

63 Non-exposed area of the neck

64 Sun-exposed area.

Immunostaining of elastin fibres: the accumulation of elastin fibres (solar elastosis) is visible (green fluorescence) in the sun-exposed area

Solar elastosis is the main histological characteristic of photo-aged skin. It consists of the accumulation of damaged elastic fibres, mainly elastin.

65 Relative elastin content in a- Sun-protected (buttock) skin in individuals of different age and sex and b- Sun-protected versus
66 moderately sun-exposed (forearm) and severely sun-exposed (facial) skin in individuals in their 60s and 70s. The relative amount of elastin was determined using computerized image analysis in immunostained sections. The values are mean ± SEM.

The results demonstrate a reduction of elastin content with age in sun-protected and sun-exposed skin associated for the latter with high elastin content, resulting in solar elastosis.

All together, these observations contribute to a better understanding of the chronological and sun-induced changes in skin elasticity and clearly demonstrate the major role of the sun regarding skin age.

Furthermore, these data were of great help for the evaluation of the efficacy of anti-aging products. For example, we have demonstrated with biomechanical and histological techniques a functional and structural remodelling of chronically sun-damaged skin after 6 months' topical treatment with ascorbic acid and madecassoside. So-called anti-ageing products should ideally combine components aiming at prevention of the environment–induced changes with agents stimulating cell metabolism, synthesis and re-structuring of the dermal fibrous matrix. However, the outcome of tissue and cell 'rejuvenation' is also partially dependent on action against already installed age-related changes. Partial removal of the damaged elements could constitute an important complementary mechanism of improvement of the physical characteristics of aged skin. Such a combined action may be attained through the local increase of antioxidants and enzyme regulating agents. Indeed, vitamin C induces collagen production in fibroblasts, partially through the modulation of collagen synthesizing enzymes.[1,2] Vitamin C also controls collagen–degrading enzymes like matrix metalloproteinase-2[3], whereas tissue concentration of ascorbic acid is significantly reduced in aged and photo-aged skin.[4,5,6]

These data are in favour of the use of topical vitamin C supplementation and we suggest that 5% ascorbic acid present in the studied formulation was at least partially responsible for the observed active restructuring of the dermal collagen and elastic tissue networks. Madecassoside, has been shown to act on the collagen synthesis in vitro[7,8,9] and in vivo[10], apparently by activating the Smad signalling pathway.[11] The association of both active components proved to be highly beneficial from the clinical, morphological and functional point of view, as demonstrated in our study.

THE OBSERVED MORPHOLOGICAL CHANGES CORROBORATE THE FUNCTIONAL EVALUATION OF SKIN ELASTICITY

Fourteen volunteers were evaluated semi-quantitatively with LM and EM. When subjects were classified into groups according to the degree of improvement in the elastic tissue structure, it turned out that 6 out of 14 were good responders, 3 were fairly good responders, whereas 5 showed no significant change. Mean percent of the functional improvement in skin elasticity (Ur/Ue cutometry values measured on the face) in the volunteers belonging to the three groups fully corroborated the morphological data and read 56.8%, 21.7%, and 4.4%, respectively.

The improvement of skin elasticity in subjects treated with vitamin C and madecassoside was noticeable in the reappearance of a normally structured superficial network.

The study showed that these components could be highly beneficial in the restructuration of the dermal collagen and elastic tissue networks. In other words, by using the right facial products you can ward off damage done by the sun to the skin.

1. Stassen FLH, Cardinale GJ, Udenfriend S. 'Activation of prolyl hydroxylase in L-929 fibroblasts by ascorbic acid'. *Proc Natl Acad Sci USA* 1973: 70: 1090–1093.
2. Murad S, Grove D, Lindberg KA, Reynolds G, Sivarajah A, Pinnell SR. 'Regulation of collagen synthesis by ascorbic acid'. *Proc Natl Acad Sci USA* 1981: 78: 2879-2882.
3. Pfeffer F, Casanueva E, Kamar J, Guerra A, Perichart O, Vadillo-Ortega F 'Modulation of 72–kilodalton type IV collagenase (Matrix Metalloproteinase-2) by ascorbic acid in cultured human amnion-derived cells'. *Biol Reprod* 1998: 59: 326–329.
4. Rhie G, Shin MH, Seo JY, et al. 'Aging- and photoaging-dependent changes of enzymic and nonenzymic antioxidants in the epidermis and dermis of human skin *in vivo*'. *J Invest Dermatol* 2001: 117: 1212–1217.
5. Leveque N, Muret P, Mary S, et al. 'Decrease in skin ascorbic acid concentration with age'. *Eur J Dermatol* 2002: 12: XXI–XXII.
6. Leveque N, Robin S, Makki S, Muret P, Rougier A, Humbert P. 'Iron and ascorbic acid concentrations in human dermis with regard to age and body sites'. *Gerontology* 2003: 49: 117–122.
7. Maquart FX, Bellon G, Gillery P, Wegrowski Y, Borel JP. 'Stimulation of collagen synthesis in fibroblast cultures by a triterpene extracted from Centella asiatica'. *Connect. Tissue Res* 1990: 24: 107–120.
8. Lu L, Ying K, Wei S, Liu Y, Lin H, Mao Y. 'Dermal fibroblast-associated gene induction by asiaticoside shown *in vitro* by DNA microarray analysis'. *Br J Dermatol* 2004: 151: 571–578.
9. Lu L, Ying K, Wei S, et al. 'Asiaticoside induction for cell-cycle progression, proliferation and collagen synthesis in human dermal fibroblasts'. *Int J Dermatol* 2004: 43: 801–807.
10. Maquart FX, Chastang F, Simeon A, Birembaut P, Gillery P, Wegrowski Y. 'Triterpenes from Centella asiatica stimulate extracellular matrix accumulation in rat experimental wounds'. *Eur J Dermatol* 1999: 9: 289–296.
11. Lee J, Jung E, Kim Y, et al. 'Asiaticoside induces human collagen I synthesis through TGFbeta receptor I kinase (TbetaRI kinase)-independent Smad signaling'. *Planta Med* 2006: 72: 324–328

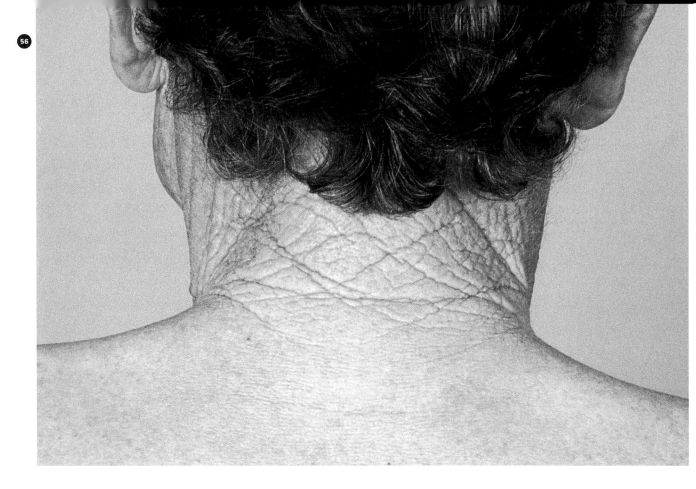

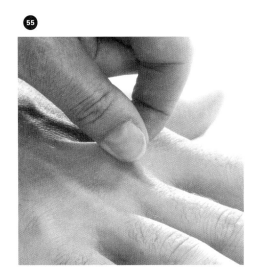

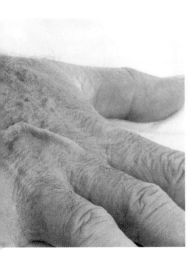

PUSH AND PULL

Elasticity	is the ability to return to the resting condition when subjected to an external force
Plants	have the ability to respond to external perturbations and to restore to their original state
Animals	also have the ability to achieve homeostasis (to control and maintain metabolism at equilibrium)
Life	is broadly defined has the ability to adapt, grow and reproduce as the external environment 'pushes' and 'pulls'
Evolution	comes about when adaptive changes are inherited and passed down from one generation to another
End Game	is death, because all things fail when they are 'pushed too far'

RUBBER-MAN

This is a portrait of the retired circus artist Michal Mozes, who was born in Poland in 1932.

He performed as a rubber-man in circuses all over the world during the nineteen sixtees including at the famous Olympia Theatre in Paris. After he almost broke his spine, which finished his career as a rubber-man, he continued working in the ring as an assistant to his wife, who was a lion tamer.

The photograph was taken in Milanowek (near Warsaw) where Michal Mozes lives with his wife. At the time this photograph was taken (11.07.2007) Michal Mozes was 75 years old and had been retired for 24 years.

Medium: c-print
Dimensions: 50×60 cm
Edition: 15 + II

SPACETIME FOAM

Space and time were for a long time the very bedrock of our thinking. In the 17th century Newton introduced absolute space and absolute time. For him, and for many scientists after Newton, they were the immutable stage on which all natural phenomena took place. According to Newton "absolute space, in its own nature, without regard to anything external, remains always similar and immovable", while "absolute, true and mathematical time, of itself, and from its own nature flows equably without regard to anything external"(*Philosophiae Naturalis Principia Mathemetica*, 1687). Nothing could be more certain and permanent.

This absolute view of the world changed dramatically with Einstein who made it all more 'relative'. He showed that space and time were an integral part of the dynamics of physics. According to the laws of his general theory of relativity, anything that carries energy and mass can curve space and time. This curvature influences the movements of bodies and manifests itself as the force of gravity. It is as if under the directorship of Einstein the stage had become alive and was promoted to be a player in the play. Later Einstein discovered that empty space should not only be considered an active element of physics, it could also carry energy by itself, just as particles and fields can.

With the advent of quantum theory even this picture had to change. As Max Planck had already realized in 1900, the nature of space changes at very small distances. At the unimaginable small scale of the Planck length, a mere $1.6 \cdot 10^{-35}$ metres, space itself loses all its meaning. At that scale the random fluctuations inherent in any quantum theory take over. One can perhaps imagine space to be like a picture on our computer screens. If one resolves the picture in greater detail, in the end the individual pixels become visible. The lines, curves and shades of the picture all disappear and only the coloured dots remain.

69 The physicist John Wheeler (1911–2008) coined the term spacetime foam to describe the fractal structures found at the very bottom of the scales of nature. Quantum space-time he imagined to be as the foam on the surface of a bath. This foam separates the air and the water. But if you look closer and closer, and see the individual soap bubbles, the difference between air and water becomes less and less clear. Where does the air stop and the bath water begin? In a similar way, according to Wheeler the notion of space and time should dissolve into quantum soap bubbles.

Wheeler's ideas have been around for some time. Only recently has his picture of the microscopic nature of spacetime become a reality within the framework of string theory. String theory is an ambitious effort to describe all particles and forces within a single unified theory of nature. Instead of point particles, the fundamental ingredients are microscopically small strings, little vibrating loops of energy. The different modes in which the string can vibrate manifest themselves as different kinds of particles, unifying these particles in an elegant geometric way. Since only one type of interaction exists—strings can split and join—the interactions between these particles are natural and completely fixed. One of the modes of the string has a special role: it describes a graviton, the fundamental quantum of gravity. Therefore string theory manages to combine the theory of relativity with the fundamental laws of quantum mechanics, one of the great outstanding puzzles of modern physics. String theory promises to be a complete theory of quantum gravity, so one should be able to see something like Wheeler's foam emerge.

70 This is indeed the case, but it happens in a surprising way. In certain models spacetime appears as a crystalline structure, as if space is made out of atoms. The size of these 'atoms' is indeed given by the Planck length, confirming the intuition of Planck and Wheeler.

This 'emergence' of space out of even more fundamental notions, inverts the absolute nature that Newton and others assigned to space and time. This emergence reflects a deeper discussion on the source of elegance and beauty in science. Here there are essentially two schools. The first works from the premise 'garbage from beauty'. From this point of view, true beauty can only be found at the most fundamental level. A prime example is the Standard Model that describes with great accuracy the known forces between the elementary particles. Its inner workings are indeed made up from very elegant mathematics, but this elegance is easily lost in the complexities of

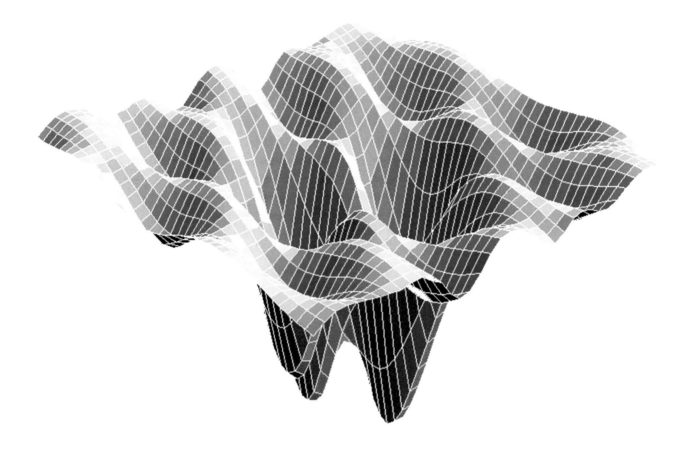

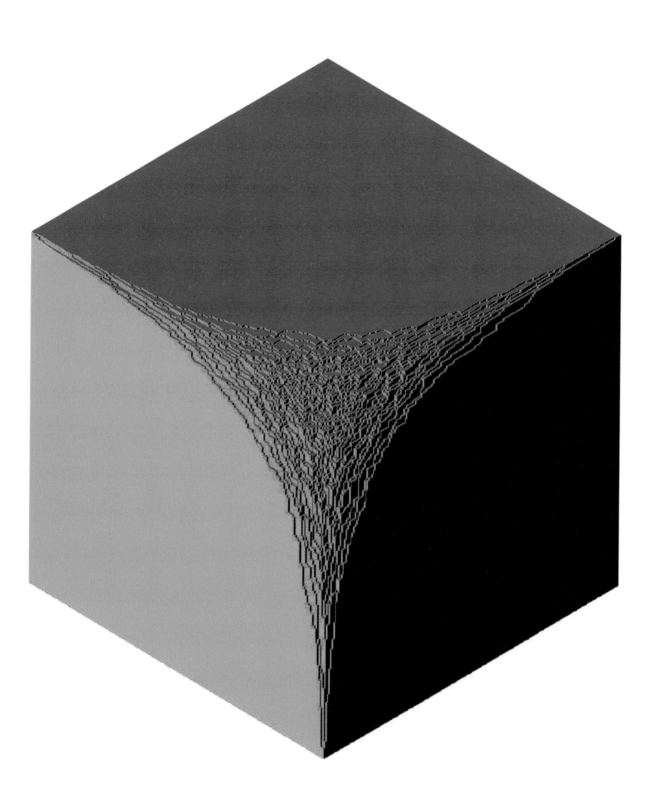

the world, the 'garbage' around us, since it is made up of uncountable numbers of these particles.

The mantra of the second school is just the opposite: "beauty from garbage". Here the leading idea is that elegance can emerge from ugly complexities. A good example is the flow of water, elegantly described by the laws of hydrodynamics. Water is a huge collection of water molecules, but in practice it is impossible to follow the interactions between them. Yet collectively they give rise to the beauty of a water wave.

One can say the newest insights about space and time seem to favour the second school, the "beauty from garbage" argument. It would be remarkable if the elegance that space and time have had for so many generations emerges from deep, ill-understood, and perhaps very ugly quantum effects.

THE SOUND OF SPACETIME

According to the definition of elasticity in physics, an object or medium is very elastic if it is very *hard* to deform. But in non-technical usage, something is elastic if it is *easy* to deform (and then springs back).

Spacetime, the stage for the dynamic drama of particles and energy fields, is incredibly elastic: it responds so stiffly that even huge amounts of energy can only create tiny perturbations. Gravitational waves are propagating oscillations of spacetime which are emitted by the rapid accelerated motion of very massive bodies, as predicted by Einstein's theory of General Relativity. Now, consider the gravitational radiation from two black holes as they merge, a process instantaneously more luminous than all the stars in the universe, and one that bathes Earth from nearby galaxies with an energy flux comparable to the full moon's. This radiation can only distort the spacetime around us by a part in 10^{-21}—the radius of an atomic nucleus compared to the radius of the Earth.

The figure shows the gravitational waveforms emitted by the in-spiral of two spinning black holes. The waveforms must be read left to right, bottom to top: as orbital energy is lost to the emission of gravitational waves, the holes spiral together and the frequency of the waves increases. In addition, the black hole spins and the orbital angular momentum interact, and tumble together, generating amplitude and phase modulations in the waves. These dynamics are described by complicated post–Newtonian equations, which extend classical gravity to include the low-velocity effects of General Relativity.

$$\frac{\dot{\omega}}{\omega^2} = \frac{96}{5}\eta(M\omega)^{5/3}\left\{1 - \frac{743+924\eta}{336}(M\omega)^{2/3} - \left(\frac{1}{12}\left[\chi_1(\hat{L}_N\cdot\hat{S}_1)\left(113\frac{m_1^2}{M^2}+75\eta\right)\right] - 4\pi\right)(M\omega)\right.$$

$$+\left(\frac{34103}{18144} + \frac{13661}{2016}\eta + \frac{59}{18}\eta^2\right)(M\omega)^{4/3} - \frac{1}{672}(4159 + 14532\eta)\pi(M\omega)^{5/3}$$

$$+\left[\left(\frac{16447322263}{139708800} - \frac{1712}{105}\gamma_E + \frac{16}{3}\pi^2\right) + \left(-\frac{273811877}{1088640} + \frac{451}{48}\pi^2 - \frac{88}{3}\hat{\theta}\right)\eta + \frac{541}{896}\eta^2 - \frac{5605}{2592}\eta^3\right.$$

$$\left.-\frac{856}{105}\log[16(M\omega)^{2/3}]\right](M\omega)^2 + \left(-\frac{4415}{4032} + \frac{661775}{12096}\eta + \frac{149789}{3024}\eta^2\right)\pi(M\omega)^{7/3}\right\},$$

71 The curves marked h_+ and $h_\times$ trace the two independent, transverse degrees of freedom of gravitational radiation, propagating toward Earth. The four overlays reprise the waves from four phases of in-spiral (shaded), and the tubes trace the oscillatory motions (grossly exaggerated) that they would impart on a ring of freely falling test particles, a useful idealization for gravitational-wave detectors. The vertical scale of the waveforms is inversely proportional to the distance from the emitting system to the observer: it is typically of order 10^{-21} for the sources that modern detectors hope to reveal. The horizontal scale is proportional to the total mass of the in-spiralling bodies: for black holes of masses comparable to our Sun, the last few oscillations would take a few milliseconds.

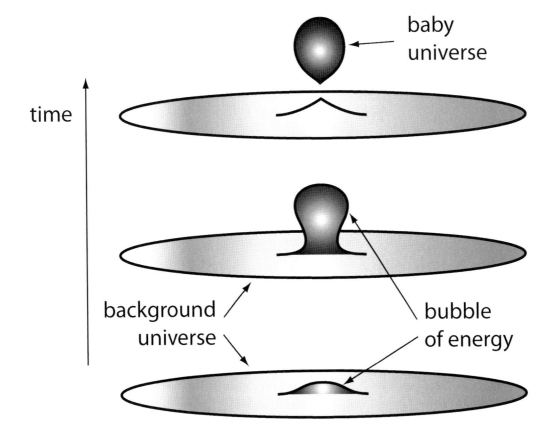

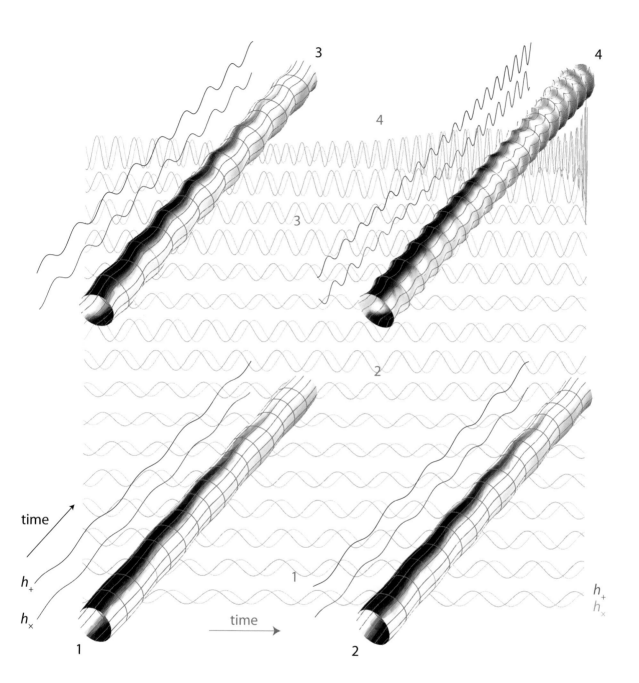

time

h_+

$h_\times$

time

h_+
$h_\times$

1

2

3

4

THE ELASTIC UNIVERSE

Albert Einstein, writing in 1954, attempted to allay the fears of anyone put off by the intimidating vocabulary of his masterwork, the general theory of relativity: "The non-mathematician is seized by a mysterious shuddering when he hears of 'four-dimensional' things, by a feeling not unlike that awakened by thoughts of the occult. And yet there is no more common-place statement than that the world in which we live is a four-dimensional spacetime continuum."[1]

In 1905, Einstein put the capstone on the special theory of relativity, the kinematic edifice that established once and for all that space and time were one. In the words of Einstein's teacher Hermann Minkowski, "Hence-forth space by itself, and time by itself, are doomed to fade away into mere shadows, and only a union of the two will preserve an independent reality."[2]

But it was ten years later than Einstein completed his greatest achieve-ment, general relativity, which took the rigid spacetime of special relativity and allowed it to have curvature, dynamics—one might say with perfect justification, elasticity.

In general relativity, spacetime can curve and warp and even come into existence or disappear. Extraordinary consequences follow from this simple idea: the irreversible disappearance of matter falling into a black hole. The Big Bang and the expansion of the universe. Different rates for clocks at different altitudes, a phenomenon that imposes itself prosaically on the satellites of the Global Positioning System.

Cosmologists are only recently coming to terms with perhaps the greatest single consequence of the elastic nature of spacetime: the ability to make a new universe. As cosy and hospitable as our local universe appears, there is a great deal about it that we don't understand. Foremost among these puzzles is the very special conditions at the beginning, near the Big Bang. It is not perhaps remarkable that the universe began at all; but given everything we know about physics, it is astonishing that its beginning was so finely-tuned and organized; low-entropy, we say in the language of thermodynamics.

We don't know why the early universe was so organized. The entropy, a measure of disorder, has been growing ever since, giving rise to every feature of the arrow of time that points from past to future. A promising scenario to explain the delicacy of our early stages is to imagine that the Big Bang was not the beginning—that there was time before the Big Bang, and that our universe was created from a prior spacetime. That our universe is, in other words, a baby, the parthenogenetic offspring of a lost parent. Modern theories of quantum mechanics suggest that a bubble of energy may fluctuate into existence, with the right properties to cause a part of spacetime to branch off and form its own universe.[3] Perhaps this special act, not of pure creation but of bending and ripping, explains the special features of the spacetime in which we find ourselves.

No one knows whether such a scenario is correct; indeed, we have not yet reached the point where we can even judge it to be coherent and sensible. But it offers a line of attack on why the early universe was such a special place. That, in turn, may help us understand the difference between the past and the future, which shows up in our everyday lives as the growth of entropy and the arrow of time.[4] And it is a scenario that could not have been even conceived were it not for Einstein's great insight that spacetime is elastic. If spacetime were fixed forever, there would be no room for the creation of new universes. But in a universe that can twist and grow and evolve, many things are possible; it's up to us to match the possible to what we actually see.

1
A. Einstein, *Relativity: The Special and General Theory.* Translated by Robert W. Lawson.: New York: Henry Holt and Company, 1920.

2
H. Minkowski, "Space and Time." Reprinted in *The Principle of Relativity*, edited by A. Einstein. New York: Dover Publications, 1952.

3
A. H. Guth, *The Inflationary Universe: The Quest for a New Theory of Cosmic Origins.* Reading: Addison-Wesley, 1997.

4
S. Carroll, *From Eternity to Here: The Quest for the Ultimate Theory of Time.* New York: Dutton, 2010.

THE ELASTIC UNIVERSE
Collage from the footage of the feature film
"Pepperminta" by Thomas Rhyner, 2009
Pierre Mennel, Camera
Ewelina Guzik, Performer

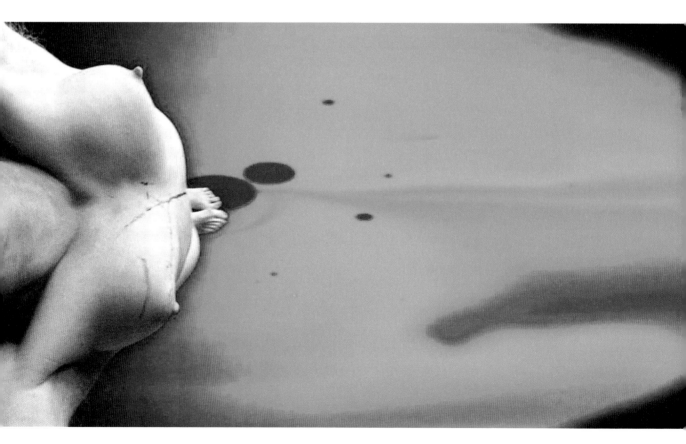

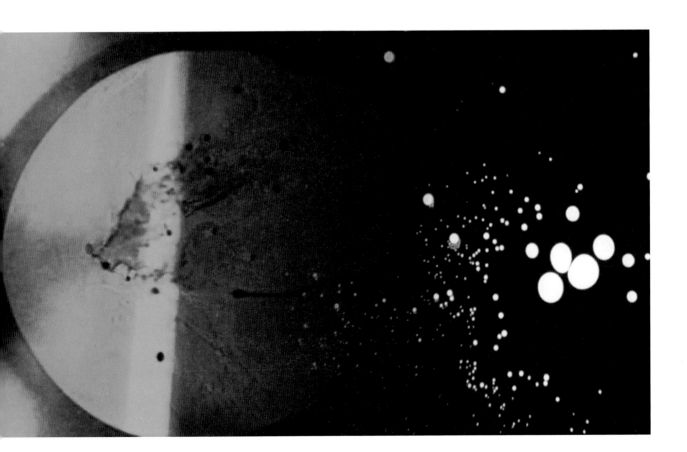

CONTRIBUTORS

Algirdas Budrevičius (Lithuania, 1954) is Associate Professor at the Department of Information and Communication in the Faculty of Communication at the University of Vilnius in Lithuania. In the past he also was the Head of the Department at the Centre for Informatics and Forecasting of the Ministry of Culture and Education and research fellow at the Institute of Scientific and Technical Information. His research focuses on cognitive science, semiotics and artificial intelligence. He has written three books. In them he presents models of cognition based on analysis of the interrelation of the phenomena of meaning and knowledge, using the dynamical systems approach. He has also proposed a general framework for analysis of the cognitive architectures of digital, neural and analogue computers and forecasts new cognitive architectures.

Andoni Luis Aduriz (Spain, 1971) studied at the Donostia School of Cuisine in San Sebastian, Spain. In 1993 he joined the team of el Bulli. Five years later he began working on his own at Mugaritz, the restaurant he has run since 1998. He has been described by the international gastronomic press as 'the most important gastronomic genius in the world of recent times'. He was awarded numerous prizes and has received two Michelin stars.
The Mugaritz restaurant collaborates with the innovation company Ibermática in order to study creativity in the restoration sector, which gave rise to the MIRAC project, a model that evaluates innovation in the field of the refined cooking. Within that framework, Aduriz is giving courses in the University of the Basque Country, since 2005.
[www.mugaritz.com]

Andrew Chapman (UK, 1971) is a designer of books, magazines and puzzles, as well as a writer and editor. He is increasingly interested in exploring puzzles and games as an art form, and has previously created an entirely lipogrammatic cryptic crossword. An inveterate generalist, he has written or contributed to books on genealogy, anthropology, scientific inventions, the internet and acting, and is the co-founder of a fringe arts festival in Oxford, UK. He has also explored the psychogeography of London's lost rivers, created a popular online book recommendation system and is currently researching decision theory.
[www.hatmandu.net]

Andrew Davidhazy (Hungary, 1941) is a member of the Imaging and Photographic Technology Department of the School of Photographic Arts and Sciences at the Rochester Institute of Technology. He has been a teacher there for over 40 years and while specializing in scientific and technical aspects of photography he is also interested in the application of technical imaging concepts to aesthetic purposes. He has lectured and exhibited worldwide and his writings have been published in numerous books, articles and journals.

Brenda Ann Kenneally (USA, 1959) is an independent journalist whose long-term projects are intimate portraits of social issues that intersect where the personal is political. She has reported on issues of class disparity between American families and for the past years her work has been a look at coming of age in post-industrial America, Upstategirls.org. She was commissioned by The New York Times Magazine to investigate the living conditions of families on The Gulf Coast on the first anniversary of Hurricane Katrina. Kenneally and independent producer Laura Loforti founded www.therawfile.org. The Raw File's mission is to publish the work of voices that have been marginalized in the traditional media and provide a theatre for open ended projects.
[www.brendakenneally.com]

Catherine Malabou (France, 1959) is a philosopher, who graduated from the École Normale Supérieure Lettres et Sciences Humaines (Fontenay-Saint-Cloud). Her aggregation and doctorate were obtained under the supervision of Jacques Derrida and Jean-Luc Marion, from the École des hautes etudes en science sociales. Her dissertation became the book, *L'Avenir de Hegel: Plasticité, Temporalité, Dialectique* (1996). She now is professor in the Philosophy Department at the Université Paris-X Nanterre and Visiting Professor in the Comparative Literature Department at the State University of New York at Buffalo. Central to her philosophy is the concept of 'plasticity', which she derives in part from the work of Hegel, as well as from medical science, for example from work on stem cells and the concept of neuroplasticity.

Chiharu Shiota (Japan, 1972) is a visual and performance artist. She has been living in Germany since 1997, where she followed the classes of Marina Abramoviç at the Hochschule für Bildende Künste, Braunschweig. In her performances and installations she is preoccupied with remembrance and oblivion. Earth and water recall the lasting and the fleeting. There is a concern with drifting between the cultures of Asia and Europe as also with a farewell to childhood. She deals generously with her dreams. Her work leads into sleep, night and the self-forgetfulness of the dreaming body. Far from Japan, her loss of orientation and her fear of losing the personal and individual have become the leitmotifs of her art. From them arise theatrical images inviting viewers participation.
Shiota has received several awards and prizes for her work and her art is acquired by international museums.
[chiharu-shiota.com]

Chris Elvin (Australia, 1955) is involved in biomaterials research and is project leader in Photoactivated Biopolymers with the Commonwealth Scientific and Industrial Research Organisation (CSIRO) and Livestock Industries in Brisbane. He also facilitates a large research project, investigating the structure and function of a near-perfect bioelastomer (resilin) found in insects. He has numerous papers published in international journals including two book chapters and is an inventor on six patents or patent applications.

Mickey Huson (Australia, 1952) is a principle research scientist at CSIRO's Division of Materials Science and Engineering. In recent years he has developed expertise in the use of Scanning Probe Microscopy (SPM), not only for acquiring images but also to examine the physical properties of a range of materials at spatial resolution of a few nanometres.

Didier Fiuza-Faustino (France, 1968) is an architect and co-founder of the label *Bureau des Mesarchitectures* in Paris. His work reciprocally summons up art from architecture and architecture from art, indistinctly using genres in a way that summarises an ethical and political attitude about the conditions for constructing a place in the socio-cultural fabric of the city. The central dimension of Faustino's line of thinking is the body. Not the body as a reference machine, but the body as a spatial component. He conceives architecture as a 'tool for exacerbating our senses and sharpening our awareness of reality'.
[mesarchitecture.com]

Eric Maskin (USA, 1950) is an economic theorist best known for his work on the theory of mechanism design. For laying the foundations of this field he shared the 2007 Nobel Memorial Prize in Economics. He has also made contributions to game theory, social choice theory, voting theory, monetary theory, contract theory, and the economics of intellectual property, among other areas. He is currently Albert O. Hirschman Professor of Social Science at the Institute for Advanced Study, Princeton.

Erik Demaine (Canada, 1981) is an associate professor in computer science at the Massachusetts Institute of Technology. His research interests range throughout algorithms, from data structures for improving web searches to the geometry of understanding how proteins fold to the computational difficulty of playing games. He received a MacArthur Fellowship as a 'computational geometer tackling and solving difficult problems related to folding and bending, moving readily between the theoretical and the playful, with a keen eye to revealing the former in the latter'. He recently published a book about folding, together with Joseph O'Rourke, called *Geometric Folding Algorithms: Linkages, Origami, Polyhedra*. Inspired by his father, he also enjoys exploring the connections between mathematics and art.
[http://erikdemaine.org]

Martin Demaine (Canada, 1942) is an artist-in-residence and visiting scientist in computer science at the Massachusetts Institute of Technology. His work crosses the borders of art and science, ranging from mathematical geometry to sculpture in paper, glass, and recycled materials. Together with his sun Erik Demaine, he co-edited *Tribute to a Mathemagician* and *A Lifetime of Puzzles* (with Tom Rodgers).

Belén Palop (Spain, 1973) is an associate professor in the Computer Science Department at the University of Valladolid in Spain. Her main area of research is computational geometry. More specifically, she has directed her attention in recent years to Voronoi diagrams, one of the most natural geometrical structures able to describe processes from forestry to crystallography or demography.

Ernesto Neto (Brazil, 1964) is a visual artist and is considered one of the leaders of Brazil's contemporary art scene. He studied at the Escola de Artes Visuais Parque Lage, and the Museu de Arte Moderna, both in Rio de Janeiro. He works with abstract installations that often take up the entire exhibition space. The installations are large, soft, biomorphic sculptures that viewers can touch, poke, and even sometimes walk on or through. He calls these environments 'Naves', which means spaceships in Portuguese. His work is acquired by several museums such as the Solomon R. Guggenheim Museum in New York and the Centre Pompidou - Musée National d Art Moderne in Paris.

Fady Joudah (USA, 1971) is a Palestinian-American poet, physician and previous field member of Doctors Without Borders. He had started writing poetry seriously in the late 1990s, 'just to get through the hell of residency'. His poetry is about those who are stateless and those who suffer. His first poetry collection, *The Earth in the Attic*, was the winner of the Yale Series for Younger Poets in 2007. He is also the translator of Mahmoud Darwish's most recent poetry collected in *The Butterfly's Burden*. In 2008 the UK's Society of Authors awarded him its prize for Arabic translation, the Saif Ghobash-Banipal award.

George Ayittey (Ghana, 1947) is a Distinguished Economist at the American University and President of the Free Africa Foundation. He is the author of *Africa* and *Indigenous African Institutions*. In these books he has championed the argument that 'Africa is poor because she is not free', that the primary cause of African poverty is less a result of the oppression and mismanagement by colonial powers, but rather a result of modern oppressive native autocrats. He also goes beyond criticism to advocate for specific ways to address the abuses of the past and present; specifically he calls for democratic government, debt re-examination, modernized infrastructure, free market economics, and free trade to promote development.

Gijs Wuite (The Netherlands, 1972) leads his group at the VU University Amsterdam focusing on exploring DNA-protein interactions and biophysical-biomechanical properties of viral capsids and cells. His work has appeared in journals and has been regularly reported in the general media. He is a recipient of a grant from the Netherlands Organization for Scientific Research and has been selected for membership of the Young Academy, part of the Royal Dutch Academy of Sciences. Exploring the synergy between art and science has been a major interest and has been the basis of his collaboration with Luciano Pinna.

Luciano Pinna (The Netherlands, 1970) is a visual artist. Constructing self-reflecting playgrounds of the real, Pinna researches the inner workings of the experience of perception. An integral part of this process is the utilization of knowledge of the cosmos and man's conscious position in it. [lucianopinna.com]

Hèctor Parra (Spain, 1976) is a composer and professor of Electroacoustic Composition at Aragon Conservatory of Music in Spain. He also is the composer-in-residence at IRCAM-Centre Georges Pompidou, in Paris. He studied in the Conservatorium of Barcelona, where he was awarded Prizes with Distinction in Composition, Piano, Harmony and Choral Direction. He holds a Master in Composition of the Paris-VIII University with honours, won numerous awards and has been commissioned by many institutions. He created a new style

of opera, a unique form where there is intercommunication between science, music and art. In this work the traditional type of opera is explored to generate a form of dramatic expression suited to the 21st century. [hectorparra.com]

Hester Bijl (The Netherlands, 1970) is a professor at the faculty of Aerospace Engineering at the University of Technology in Delft. Her work contains simulations of currents around aeroplane wings and wind turbine blades. She researches the influence of uncertainties of, among others, material characteristics.

Sander van Zuijlen (The Netherlands, 1967) is an aerospace engineer and assistant professor at the faculty of Aerospace Engineering at the University of Technology in Delft where he got his Masters degree in 2001 on the subject of 'Direct numerical simulation of 1d, unsteady, compressible, viscous flow'. Together with Hester Bijl he specializes in the topic of numerical simulation of fluid-structure interaction, and coupling of different mathematical systems and computational aero-elasticity. In 2006 he successfully defended his PhD thesis entitled Fluid-Structure Interaction Simulation —Efficient higher order time integration of partitioned systems.

Ioana Ieronim (Romania, 1947) is a poet, translator and playwright. She has been flexible about poetic form, writing for trilingual publication and video poetry, as well as producing two memoirs in verse. Both these books recorded the lost Saxon communities of mid-century Transylvania. 'Found' elements, whether of folk iconography or from the era of digital imaging, feature as structural ideas. As former cultural attaché in Washington, she has directed the Fulbright programme in Romania. Her work has been widely translated. Currently she is a member of the Writers' Guild in Romania and of the Romanian PEN-Club.

Ionat Zurr (UK, 1970) **and Oron Catts** (Finland, 1967) are artists, researchers and curators, who formed the Tissue Culture and Art Project, an ongoing research and development project into the use of tissue culture and tissue engineering as a medium for artistic expression. They have been artists in residence in the School of Anatomy and Human Biology since 1996 and were central to the establishment of SymbioticA, the Centre of Excellence in Biological Arts at the University of Western Australia. Catts is the co-founder and director of SymbioticA and Dr. Ionat Zurr is researcher and SymbioticA's academic coordinator. They have recently exhibited as part of the Medicine and Art Exhibition at the Mori Museum in Tokyo and the Design and Elastic Mind Exhibition in the Museum of Modern Art in New York. [tca.uwa.edu.au]

Jim Gimzewski (Scotland, 1951) is a professor of Chemistry at the University of California and member of the California NanoSystems Institute. He pioneered research on mechanical and electrical contacts with single atoms and molecules using scanning tunneling microscopy (STM) and was one of the first people to image molecules with STM. His approach was recently used to convert biochemical

recognition into nanomechanics. His current interests are in the nanomechanics and elasticity of cells and bacteria where he collaborates with the UCLA Medical and Dental Schools. He is involved in projects that range from the operation of X-rays, ions and nuclear fusion using pyroelectric crystals and single molecule DNA profiling. He is also involved in numerous art-science collaborative projects that have been exhibited in museums throughout the world. As co-director of the UCLA Art|Sci Center, he has promulgated the fusion of artistic creation with scientific innovation.

Karl Niklas (USA, 1948) is the author of numerous peer-reviewed scientific papers and three books: *Plant Biomechanics, Plant Allometry*, and *The Evolutionary Biology of Plants*. Currently he is the Liberty Hyde Bailey Professor of Plant Biology in the Department of Plant Biology, Cornell University, where he teaches courses in basic plant biology and evolutionary biology. Niklas is a past president of the Botanical Society and a past Editor-in-Chief of the American Journal of Botany. He is the recipient of numerous awards, including a John Simon Guggenheim fellowship and the Alexander von Humboldt Stiftung Prize for senior American scientists.

Kevin Carlsmith (USA, 1967) is an associate professor of psychology at Colgate University. In his work he studies the psychological motivations that drive the desire for punishment in biological and interpersonal contexts. His current projects include experiments on how personal bias influences formal sanctions, and US attitudes about torture and aggressive interrogation practices. His research is funded by the National Science Foundation, and he is a fellow at the Center for Advanced Study in Behavioral Sciences.

Kristin Henrard (Belgium, 1970) is a professor in human rights and minority protection and associate professor in constitutional law at Erasmus University, Rotterdam. She worked at the Constitutional Court of South Africa as researcher for Judge Kriegler and monitored the South African constitutional negotiations in 1996 for the Flemish Government. Her main publications pertain to the areas of human rights and minority protection. She is also managing editor of the Netherlands International Law Review, a member of the international advisory board of the Global Review of Ethnopolitics, a country specialist on South Africa for Amnesty International and a member of the Young (KNAW).

Kwon Won Tae (Republic of Korea, 1967) is a traditional Korean tightrope walker who devoted 34 years of his life to walking and dancing on a tightrope. Tightrope walking, or eoreum, comes from the Korean word for ice, which is an indication of how difficult the practice is. The skill has recently experienced a renaissance in the film *King and the Clown*, which featured a troupe of entertainers who became court jesters. In the movie, Kwon starred in tightrope walking scenes. He has completed the NamsadangNori (all-male vagabond clown theatre), which was given UNESCO Important Intangible Cultural Heritage status in 2009. He participated in many international competitions and festivals and is touring through Asia.

ana Šlezić (Canada, 1973) is a photographer who believes in cohesive bodies of work that communicate an issue or concept. She left the steadiness of the media and began her own projects, where she could discover the art of storytelling. For two years she has documented the life of women in Afghanistan, about which she published her first book *Forsaken*. Her work has been awarded with numerous prizes, including the World Press Photo Award in the Portrait Story category. Currently, Šlezić is based in New Delhi. [anaslezic.com]

Lars Spuybroek (The Netherlands, 1959) is the principal of NOX, an architecture and art studio in Rotterdam. Since the early 1990s he has been researching the relationship between art, architecture and computing. He received international recognition after building the Water Pavilion in 1997 (H2OtwoExpo), the first building in the world fully incorporating new media. NOX completed the interactive D-Tower and the Son-O-house, plus a cluster of cultural buildings in Lille, France (Maison Folies). The first fully theoretical account of his work, titled *The Architecture of Continuity*, was published with V2_NAI publishers in 2008. Lars Spuybroek has won several prizes and has exhibited all over the world. Since 2006 he has been a full professor and the Ventulett Distinguished Chair at the Georgia Institute of Technology in Atlanta. [nox-art-architecture.com]

Margriet Sitskoorn (The Netherlands, 1966) a neuropsychologist and Professor of Clinical Neuropsychology at the Faculty of Social and Behavioural Sciences at the University of Tilburg. She focuses on the question of how we can develop our brain by changing our behaviour and surroundings. In her book *The Mutable Brain* she writes about this matter. She also is the Director of the Neurocognitief Centrum Nederland. Apart from her scientific work Sitskoorn also writes columns for different popular scientific magazines and collaborates in making television programmes.

Marjorie Schick (USA, 1941) is an artist who makes jewellery objects and body sculptures. Her work is a statement which is complete when off the figure, yet is constructed and exists because of the human body. She is intrigued by the idea that the human body is capable of carrying large objects, both physically and visually. There are five major aspects to her work: the constructed three-dimensional form, colour relationships, the definition of space, the combination of patterns, and the scale of the objects in relationship to the human figure. She creates a sense of visual tension among the formal elements of each object. Her work has been exhibited in museums and galleries around the world and added to several museum collections.

Mark Peletier (UK, 1969) studied mathematics in Leiden and Paris, where he specialized in non-linear differential equations. After his masters degree he continued with a PhD research project in Delft and at the Centrum voor Wiskunde en Informatica in Amsterdam. The subject of the research was degenerate diffusion, a type of diffusive transport that appears in various applications, such as pollutant transport in groundwater. During a postdoc period in Bath he developed a further interest in the relationship between mathematical structures on one hand and the real-world systems that they describe on the other. This deep relationship is often only weakly understood, and gives rise to beautiful problems in both pure and applied mathematics. He is also interested in the societal side of mathematics and science in general, contributing to public understanding of science via projects such as The Eurodiffusion Experiment.

Medhi Walerski (France, 1979) had an education in classical ballet, modern dance and improvisation. After he joined the Paris Opéra Ballet, the Nice Opéra Ballet and Ballet du Rhin as a professional dancer, he exchanged France for the Netherlands and continued his dance career with Nederlands Dans Theater in The Hague. Besides being a dancer, Walerski is also a choreographer and made productions as *Moume, Buried Dead or Alive* and *Mammatus*. In December 2007 he contributed to 'Sharing Art', a project for street children in Bangladesh initiated by choreographers duo Lightfoot León. Currently Walerski works worldwide as a freelance dancer and choreographer and has been commissioned to create new work for the Bern Ballet and Götenberg Ballet.

Michele Vallisneri (Italy, 1973) works as theoretical physicist at NASA's Jet Propulsion Laboratory. He is a member of the LIGO Scientific Collaboration and is U.S. Deputy Mission Scientist for LISA, a planned space-based gravitational-wave observatory. His research interests span the detection, analysis and interpretation of gravitational-wave signals, computational physics, and the creative interface of science and art, as explored through music, visualization and computer programs. He is listed as author of numerous articles in international scientific journals, and is a co-author of *Einstein's Cosmic Messengers,* a multimedia concert about gravitational waves [andreacentazzo.com/ecm]

Minsuk Cho (South Korea, 1966) graduated from the Architectural Engineering Department of Yonsei University in Seoul and the Graduate School of Architecture at Columbia University in New York. He began his professional career working for Kolatan/MacDonald Studio, and Polshek and Partners in New York; later he moved to the Netherlands to work for OMA. Through these jobs, he gained experience in a wide range of architectural and urban projects implemented in various locations. With partner James Slade, he established Cho Slade Architecture in 1998 in New York City, to be engaged in various projects both in the U.S. and Korea. In 2003, he returned to Korea to open his own firm, Mass Studies, a critical investigation of architecture in the context of mass production, intensely over-populated urban conditions, and other emergent cultural niches that define contemporary society. [massstudies.com]

Min Xiao-Fen (China, 1961) is a pipa player and vocalist, known for her work in traditional Chinese music, contemporary classical music and jazz. She studied with her father, Min Ji-Qian, a professor and pipa instructor at Nanjing University and performed as pipa soloist for the Nanjing National Music Orchestra from 1980. In 1992 she immigrated to the United States and since that time has worked with numerous contemporary composers, jazz saxophonists and the singer/songwriter Björk. She is also the founder of Blue Pipa Inc., a non-profit organization dedicated to the exploration of traditional and modern music from all cultures. [bluepipa.org]

Nick White (UK, 1955) studied science and received a PhD in chemistry at the University of Bristol. He works as Technical Risk Director at SSL International; a focused consumer brand company and owner of the leading global brands Durex and Scholl. Since 1996 he has been a participant in UK, European and International standardization efforts for condoms. He published his research paper, *Contraception*, on the mechanism of breakage of male condoms, based upon his work at SSL, showing that nearly all condoms break by a 'blunt puncture' mechanism.

Odmaa Bayartsogt (Mongolia, 1986) is a contortionist. She started contortion when she was eight years old and practiced for two years before becoming a professional artist and performing in Asia and Europe for seven years. Since 2005 she has been with 'O' at Cirque du Soleil.

Pernille Fischer Christensen (Denmark, 1969) is a film director, who also writes her movies, as well as acting in them. She graduated from the National Film School of Denmark with the movie *India*, which later went on to win an award at the Film Festival Cannes. Her first feature film, *A Soap*, centred on the relationship between a female beauty shop owner and the transsexual living downstairs. The movie opened at the Berlin Film Festival and has won several awards. Her second feature film called *Dancers*, tells the story of a dance school run by the bright and lively Annika and her no-nonsense mother. Her multi-film preoccupation deals mostly with distraught female characters who get misled by their out-of-control emotions and land in relationships with ill-advised romantic partners.

Pipilotti Rist (Switzerland, 1962) is a video-artist. Her focus is on video/audio installations because there is room in them for everything like painting, technology, language, music, movement, poetry etc. Her opinion is: 'Art's task is to contribute to evolution, to encourage the mind, to guarantee a detached view of social changes, to conjure up positive energies, to create sensuousness, to reconcile reason and instinct, to research possibilities and to destroy clichés and prejudices.' Her works have been exhibited widely at museums and festivals throughout Europe, Japan and the US. Her multimedia work blurs the boundaries between visual art and popular culture and explores the unfamiliar in the everyday. [pipilottirist.net]

Thomas Rhyner (Switzerland, 1961) is a graphic designer and art director for different advertising agencies in Switzerland. His design work spans the areas of theatre, art and books.

Rafal Milach (Poland, 1978) works as a freelance photographer for *Newsweek Poland, Polityka* and *Przekroj magazine*. Aside of his editorial assignments Rafal has been working on personal projects such as: *The Grey, Disappearing Circus, Ukraine by the Back Sea* and *Young Russia*. These projects concentrate on the most industrial and ecologically devastated regions of Europe and youth in post Soviet Russia. His photos are internationally awarded and exhibited. Rafal is also a member of Austrian based Anzenberger Agency. [rafalmilach.com]

Rajesh Mehta (India, 1964) was raised in the United States and has been professionally based in Europe and presently in Asia where he is the artistic director and founder of ORKA-M: International Institute of Innovative Music. He is an award-winning trumpet player, composer, producer and engineer. His extended musical vision has created new instrumental designs such as his 'hybrid trumpet' as well as the music-architecture project series 'Sounding Buildings' in which he employs his innovative notational framework of graphical drawings called 'imaginational maps'. [rajesh-mehta.com]

Robbert Dijkgraaf (The Netherlands, 1960) is Distinguished University Professor of Mathematical Physics at the University of Amsterdam and President of the Royal Netherlands Academy of Arts and Sciences (KNAW). As a member of a research group he examines string theory, quantum gravity and the interface of mathematics and particle physics. His interests continue in creating more public awareness of mathematics and science, and bridging the gap with the arts and humanities, which he writes about in columns. He is recipient of the Physica Prize of the Dutch Physical Society and the NWO Spinoza Prize.

Rodger Kram (USA, 1961) has earned PhD degrees in biology and biomechanics at Northwestern, Penn State and Harvard Universities. He studies and admires how humans and a wide variety of animals walk and run. These animals have included ants, rhinoceros beetles, dogs, llamas, horses, alligators, crocodiles, Galapagos tortoises, elephants and, of course, kangaroos. He is an associate professor in the Integrative Physiology department at the University of Colorado, Boulder USA. He lives, walks and runs at an elevation of 2400m in the Rocky Mountains.

Sean Carroll (USA, 1966) is a theoretical physicist at the California Institute of Technology. His research involves theoretical physics and astrophysics, focusing on issues in cosmology, field theory, and gravitation. He is the author of the upcoming *From Eternity to Here*, a book on the arrow of time, and *Spacetime and Geometry*, a book on general relativity. He also has produced a set of introductory lectures for The Teaching Company entitled *Dark Matter* and *Dark Energy: The Dark Side of the Universe*.

Shfaqat Abbas Khan (Pakistan, 1972) studied geodesy at the University of Copenhagen. Since 2007 he has held a research position at the National Space Institute in Copenhagen (DTU Space). From 2005 to 2007 he was employed as a Research Scientist at the Department of Geodesy, Danish National Space Center, Denmark. He has served as Principal Investigator or Co-Investigator in several glaciological studies of Greenland. In 2005 he received the Young Authors' Award from the International Association of Geodesy.

Sonia Cillari (Italy, 1970) is a media artist and architect who lives and works in Amsterdam. Her work involves the creation of sensorial and perceptual mechanisms in immersive and augmented environments. Her artistic investigation examines how patterns of consciousness, perception and identity emerge in such settings. Over the last years she has been specifically interested in a field of research concerning the 'Body as Interface'. Her interactive installations, at the intersection of architecture and performance art, have been exhibited and presented internationally and have won numerous prizes. Since 2007, she has been teaching media art at the Frank Mohr Institute, Interactive Media Environment Department in Groningen. [soniacillari.net]

Sophie Seité (France, 1960) joined the Pharmacological and Photobiology group in L'Oréal Life Science Research laboratories in Paris, where she conducted research on the clinical, biophysical and histological effects of acute and repetitive exposure to UV light for over 20 years. She also evaluated the safety and efficacy of new molecules and finished products, mainly in the fields of ageing, photo ageing and photo protection. Since 2005, she has managed international clinical studies performed on products for dermatologists.

Catherine Gerst (France, 1960) was a researcher at the International Centre of Dermatological Research in Sophia-Antipolis where she studied the expression of genes regulated by retinoids. Later she joined the L'Oréal Advanced Research division where she conducted research on hair and skin biology and genetics. Since 2001 she has been in charge of scientific communication within the L'Oréal Research and Development division.

Sozyone Gonzalez (Belgium, 1973) was born by the name of Pablo Gonzalez. He attended the BrusselsAcademy of Fine Arts from 1990 to 1996. During that period he formed a great interest in metropolitan graffiti and aesthetic vandalism, and officially became Sozyone. In 1996, he developed with other artists a new form of graffiti, a brutally refined mix of Marvel Comics, abstract futuristic mathematics, alphabetical constructivism and facial cubism: pure graffiti vanguard, that dismisses any other forms of graffiti as futile. Since 2004 Sozyone has exposed his work, which is sought by certain purists, and became Sozyone Gonzalez for his new public. [sozyone.com]

Theodore Eliades (Greece, 1965) is an Associate Professor at the Aristotle University of Thessaloniki and affiliated with Texas, Marquette, Manchester, and Bonn universities. He graduated from the University of Athens and the postgraduate orthodontic program of the Ohio State University and holds degrees in biomaterials. He has published numerous articles and book chapters and is the founding editor of several orthodontic journals. The diffusion of this work into fields associated with natural and engineering sciences led to a full membership in both the Royal Society of Chemistry and the Institute of Physics.

Thomas Hesselberg (Denmark, 1975) got his Master of Science in Zoology from the University of Aarhus in 2002 on a project of web-building in drugged spiders. He then moved to the United Kingdom, where he in 2006 received his PhD from the University of Bath on a biomimetic project on locomotion and functional morphology of ragworms. He is currently a Marie Curie fellow at the University of Oxford, where he works on biomechanical aspects of wind loading and prey impact in orb webs, but has also worked on web-building behaviour in tropical spiders at the Smithsonian Tropical Research Institute in Panama and on insect flight at the University of Ulm in Germany. In addition to his research, he is involved with science communication and consultancy through the Danish company ZenSci Consult. [thomashesselberg.com]

Colophon

Editors: Hester Aardse & Astrid van Baalen
Copy editor: Andrew Chapman
Design: studio Joost Grootens / Joost Grootens
with Annemarie van den Berg, Barbara Hoffmann,
Tine van Wel, Manuel Wesely
Lithography: Pieter van der Meer
Printing: Lecturis, Eindhoven
Publisher: Lars Müller Publishers
Printed on Gardapat 13 Kiara, 115 gr/m²
Cover: Invercote G, 350 gr/m²

Acknowledgements
Our gratitude to the following people without whom
this book would not be possible:
All the contributors and guest advisors, Jonne Verburg,
Ken Arnold at the Wellcome Trust, Sanghee Song,
Maarten Asscher, Dave Sands, Simon Wragg, Amanda
Pinatih and Marijke Evers.

This publication would not have been possible
without the support of
SNS REAAL Fonds
Overvoorde-Gordon Stichting / Pauwhof Fonds
Elise Mathilde fonds
Mont Blanc *Creative Assistance Group*
Cartiere del Garda
Proost en Brandt Papier

Pars Foundation
Amsterdam, The Netherlands
www.parsfoundation.com

Lars Müller Publishers
5400 Baden, Switzerland
www.lars-mueller-publishers.com

Findings on Elasticity is the second volume in
Pars' Atlas of Creative Thinking. The first volume is
Findings on Ice

ISBN 978-3-03778-148-7

Sources / credits

Andoni Luis Aduriz
translation Jennifer Farmer

Nick White
photos: Andrew Davidhazy

Chiharu Shiota
photos: Sunhi Mang / copyright: Chiharu
Shiota

Pipilotti Rist
For complete credits see:
www.pepperminta.com
Courtesy the artists and Hauser & Wirth

Odmaa Bayartsogt
Also pictured in the photos are: Saraana
Gantumur (under Odmaa); Namchinkhand
Damba-Kaye (bottom row on the left);
Enkhjargal Dashbaljir (bottom row on the
right). "O" by Cirque du Soleil at Bellagio,
Las Vegas.
Photo by Richard Termine/ Costumes by
Dominique Lemieux / ©Cirque du Soleil Inc.

Medhi Walerski
translation Marijke Mayer

Chris Elvin
Dragonfly image 40 by David McClenaghan,
CSIRO Canberra, Australia. Layout by Dr
Nancy Liyou and Ted Hagemeijer (Brisbane,
Australia).
Image 41 by Dr David Merritt and Darren
Wong, University of Queensland).

Scarlet
Photo: Stephen Woods
Courtesy: newstream.co.uk
Photographs taken on Airedale Farm in
Shropshire, England.

SNS REAAL Fonds